LIST OF TITLES

Already published

Cell Differentiation — J.
Biochemical Genetics — R..
Functions of Biological Membranes — M. Davies
Cellular Development — D. Garrod
Brain Biochemistry — H.S. Bachelard
Immunochemistry — M.W. Steward
The Selectivity of Drugs — A. Albert
Biomechanics — R. McN. Alexander
Molecular Virology — T.H. Pennington, D.A. Ritchie
Hormone Action — A. Malkinson
Cellular Recognition — M.F. Greaves
Cytogenetics of Man and other Animals — A. McDermott
RNA Biosynthesis — R.H. Burdon
Protein Biosynthesis — A.E. Smith
Biological Energy Conservation — C. Jones
Control of Enzyme Activity — P. Cohen
Metabolic Regulation — R. Denton, C.I. Pogson
Plant Cytogenetics — D.M. Moore
Population Genetics — L.M. Cook
Insect Biochemistry — H.H. Rees
A Biochemical Approach to Nutrition — R.A. Freedland, S. Briggs
Enzyme Kinetics — P.C. Engel
Polysaccharide Shapes — D.A. Rees
Transport Phenomena in Plants — D.A. Baker
Cellular Degradative Processes — R.T. Dean
Human Genetics — J.H. Edwards
Human Evolution — B.A. Wood

In preparation

The Cell Cycle — S. Shall
Microbial Metabolism — H. Dalton, R.R. Eady
Bacterial Taxonomy — D. Jones, M. Goodfellow
Metals in Biochemistry — P.M. Harrison, R. Hoare
Muscle — R.M. Simmons
Xenobiotics — D.V. Parke
Biochemical Systematics — J.B. Harbourne
Membrane Assembly — J. Haslam
Isoenzymes in Biology — C. Taylor, R. Rider
Invertebrate Nervous Systems — G. Lunt
Genetic Engineering: Cloning DNA Molecules — D. Glover

OUTLINE STUDIES IN BIOLOGY

Editor's Foreword

The student of biological science in his final years as an undergraduate and his first years as a graduate is expected to gain some familiarity with current research at the frontiers of his discipline. New research work is published in a perplexing diversity of publications and is inevitably concerned with the minutiae of the subject. The sheer number of research journals and papers also causes confusion and difficulties of assimilation. Review articles usually presuppose a background knowledge of the field and are inevitably rather restricted in scope. There is thus a need for short but authoritative introductions to those areas of modern biological research which are either not dealt with in standard introductory textbooks or are not dealt with in sufficient detail to enable the student to go on from them to read scholarly reviews with profit. This series of books is designed to satisfy this need. The authors have been asked to produce a brief outline of their subject assuming that their readers will have read and remembered much of a standard introductory textbook of biology. This outline then sets out to provide by building on this basis, the conceptual framework within which modern research work is progressing and aims to give the reader an indication of the problems, both conceptual and practical, which must be overcome if progress is to be maintained. We hope that students will go on to read the more detailed reviews and articles to which reference is made with a greater insight and understanding of how they fit into the overall scheme of modern research effort and may thus be helped to choose where to make their own contribution to this effort. These books are guidebooks, not textbooks. Modern research pays scant regard for the academic divisions into which biological teaching and introductory textbooks must, to a certain extent, be divided. We have thus concentrated in this series on providing guides to those areas which fall between, or which involve, several different academic disciplines. It is here that the gap between the textbook and the research paper is widest and where the need for guidance is greatest. In so doing we hope to have extended or supplemented but not supplanted main texts, and to have given students assistance in seeing how modern biological research is progressing, while at the same time providing a foundation for self help in the achievement of successful examination results.

J.M. Ashworth, Professor of Biology, University of Essex

Motility of Living Cells

P. Cappuccinelli

Professor of Microbiology,
Institute of Microbiology,
School of Medicine,
University of Sassari,
Sardinia, Italy

Chapman and Hall
London and New York

First published in 1980 by
Chapman and Hall Ltd
11 New Fetter Lane, London EC4P 4EE
Published in the USA by
Chapman and Hall Ltd
in association with Methuen, Inc.
733 Third Avenue, New York, NY 10017

Printed in Great Britain at the
University Printing House, Cambridge

ISBN 0 412 157770 5

British Library Cataloguing in Publication Data

Cappuccinelli, P
Motility of living cells.—(Outline studies in biology).
1. Cells—Motility
I. Title II. Series
574.8′764 QH647 79–41113

ISBN 0-412-157770-5

Contents

Acknowledgement

Special thanks are to be given to my wife Maria for the beautiful figures, to my friends Ted Weinert and Salvatore Rubino for help in manuscript preparation, and to my dog Kelly for not having eaten the manuscript during the long stage of its assembly. I hope this is not due to a general indigestibility of the content.

1 Introduction

Philosophers through the ages have made the astute observation that life, in its many aspects, appears to be continuously moving. All things in the universe, from the cosmic to the atomic level, exhibit some form of movement. Getting down to earth, the capacity to move is also an essential feature of the biological world.

Movement was in fact synonymous with life at the time that Antony van Leeuwenhoek made the first simple microscope (seventeenth century). Using his primitive instrument he observed micro-organisms, which he called 'animalcules', swimming through a drop of water, and therefore he proclaimed that they 'seem to be alive'.

We now know that movement in this form is not a prerequisite for life, although it is a crucial aspect in many living organisms.

Realizing the general importance of motility in the biological world, this book will try to focus on the motility at a cellular level. Motility at a cellular level can take one of several forms: movement of components within the cell itself or movement of the cell as a whole (cell locomotion). In prokaryotic cells, motility phenomena are restricted largely to the movement of the cell as a whole, since intracellular transport appears to be much simpler than in eukaryotes. The presence of a more complex system of cytoplasmic organelles in eukaryotic cells, together with their need for efficient control of membrane behaviour, necessitates a more complicated intracellular motility network. These higher cells have also developed effective systems for the motility of the cell as a whole. Cell locomotion is, in fact, necessary both for giving an individual cell the capacity to respond to its environment and for enabling cell interaction in multicellular organisms. Despite the differences between these two types of motility, the underlying molecular mechanisms are basically the same.

One crucial feature of locomotion in all cells is its directionality. Cells are in fact able to sense changes in their environment and respond accordingly by regulating the directionality of their movement. Such a capacity to react actively to environmental stimulation gives to the cell, amongst other things, nutritional advantages as well as the possibility of social life.

Although cell motility is one of the oldest fields in scientific research, the knowledge of molecular mechanisms has increased particularly in recent years as a consequence of the advances in genetic and biochemical techniques. The exact nature of many of the molecular mechanisms involved remains controversial, especially in the area of the control of motility. However, there is good reason to expect that the intensive research now going on in this area will be able to resolve many of these problems in the next few years.

In the next chapter of this book we will discuss the motility of prokaryotes and particularly the flagellar-dependent motility of bacteria. Other interesting aspects of motility in simple organisms, such as the gliding movement of myxobacteria or leptospiral motility, cannot be treated here. The following chapters will deal with the motility systems of eukaryotes. Emphasis will be placed on results from current research on the molecular aspects of the problems.

Although each topic cannot be treated in depth, at the end of each chapter the reader will find a list of references from which he can gain a more thorough understanding of the subject.

Topics for further reading

Dobel, C. (1932), *Antony van Leeuwenhoek and his 'Little Animals'*, Constable, London. Reprinted in paperback by Dover, New York (1960).

A collection of A. van Leeuwenhoek's letters to the Royal Society of London, in which he describes, in a charming style, the observations made with his microscope.

Interesting reviews on the different types of cell movement and the various ways it is generated may be found in:

Aspects of Cell Motility (1968), *Symp. Soc. Exp. Biol.*, **XXII**, Cambridge University Press, Cambridge.

Molecules and cell movement (1975) (ed. S. Inoué and R.F. Stephens), Raven Press, New York.

2 Motility in prokaryotic organisms

Scientists have long been fascinated by motility in simple organisms such as bacteria. Bacteria are indeed an attractive system for the study of motility and its related problems because they have a relatively simple cell organization (in comparison with other cell types such as eukaryotic cells), their biochemistry has been studied in depth, and furthermore, due to their ability to reproduce rapidly and in vast numbers, they provide an ideal system in which to use genetic techniques for studying biological problems.

The combination of biochemical and genetic approaches has proved fruitful for the study of motility at a molecular level and of the structures responsible for cell motility; and the results so far are very promising.

Bacteria have not evolved different types of motility as have eukaryotic cells; their motility always depends on the presence of a structure called a flagellum which is able to rotate like a propeller and thus move the cell.

The majority of bacterial flagella are attached at one end to the cell, leaving the rest of the flagellum free. In some bacteria, however, (e.g., spirochetes) the complete length of the flagellum is attached to the cell body by an external membrane. In both cases cell motility arises from the rotatory movement of the flagellum.

Bacteria are able to control this movement. They can sense and record the concentration of substances with which they come into contact and they can then respond accordingly. This ability to record information can be considered as a primitive model of a memory system.

Although cell movement is not indispensable for the existence of the cell, and non-motile strains can grow and multiply, motility is widespread amongst bacteria and especially amongst free-living forms. In particular situations, motility can be of advantage to bacteria, allowing them to escape from a toxic environment or to move towards a source of food.

In the following pages we shall examine the characteristics of bacterial flagella from a morphological and biochemical point of view. We shall also discuss how synthesis of the flagella structure is genetically controlled and how the chemotactic system functions and orientates bacterial movement.

2.1 Elements involved in motility

Bacterial flagella are long and slender appendages protruding from the bacterial cell. They can be found singly or in a group at one or both poles of the cell, or they can completely cover the external surface, giving a hairy appearance to the bacterium (fig. 2. 1). This pattern is a genetically stable characteristic and has been used to classify bacteria into two orders: *Eubacteriales,* showing a random flagellar distribution, and *Pseudomonadales,* showing a polar flagellar distribution.

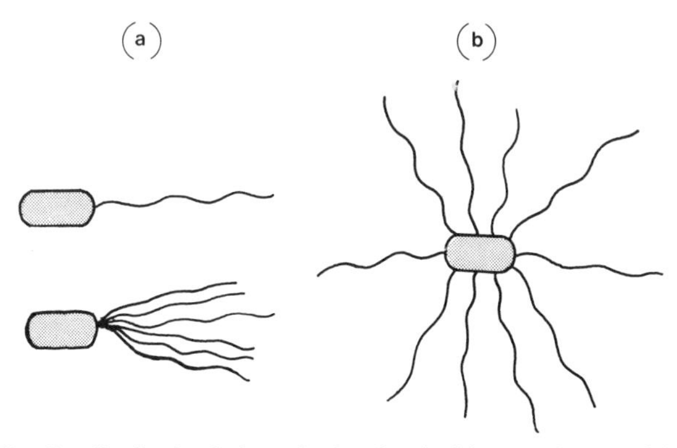

Fig. 2.1 Flagellar distribution in bacteria showing, in (a), polar flagellated bacteria with a single flagellum (top) or a bundle of flagella (bottom) originating from one pole of the cell, and in (b), a peritrichous bacterium with flagella distributed over the entire cell surface.

Due to their narrowness (120–250 nm diameter) flagella can hardly be seen with standard microscopical methods. However, using appropriate techniques, they can be studied. For example, flagella can be stained with a solution containing a precipitating agent, such as tannic acid, which forms a deposit around the flagellum increasing its width and allowing it to be examined with a normal microscope. Using dark field microscopy, they can be even more clearly defined, appearing as distinct lines of light against a black background. Under these conditions they can be seen as thin filaments, normally longer than the cell from which they originate. In the rhodobacterium *Chromomatium okenii,* a large cell (9–15 μm) often employed in early studies on bacterial motility, the flagella have been found to originate from one of the poles in a group of about 40 filaments, each approximately 25 μm long. In the well-known enteric bacterium *E.coli,* there are usually 6 flagella, each 2 to 3 times longer than the bacterial cell.

Another important morphological feature that emerges from the use of the microscope is the shape of the flagella. They always have a wave-like or helical appearance, characteristic of all bacterial strains, and protrude from the cell at irregular angles, following no fixed pattern.

A more precise definition of the morphology of the flagella can be obtained using the electron microscope. The classical work done by De Pamphilis and Adler [35, 36] and Dimmit and Simon [38] with *E.coli* and *B.subtilis* has not only clarified the fine aspects of flagellar ultrastructure, but has also provided the basis for understanding the mechanism of bacterial motility. These authors were able to set up a technique for obtaining large amounts of intact flagella. This technique utilizes the enzymatic digestion of the cell wall, cell membrane lysis with detergent and isolation by differential and gradient centrifugation in CsC1 away from cell debris. With this method it is possible, through the removal of the bacterial envelopes as described above, to obtain the complete flagellum, including all the structures that attach the flagellum to its cell. Examination of this material with the electron microscope has clarified many structural aspects of the flagellum. Basically, it can be divided into three district substructures: the filament, the hook, and the basal body. The first two are found outside the cell envelopes and the third lies between them. A normal *E.coli* filament has a helical shape, a wavelength of 2.3 μm, and a diameter of 20 nm, but mutants have been isolated with markedly different characteristics (Fig. 2.2). For example, *straight mutants* lack the helical shape almost completely and *curly mutants* form, as a stable characteristic, filaments with tighter helical waves

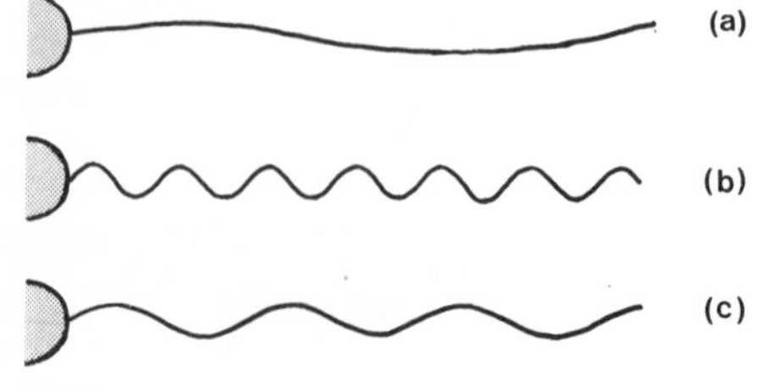

Fig. 2.2 Patterns of flagellar undulation: (a) = straight mutant; (b) = curly mutant; (c) = normal flagellum.

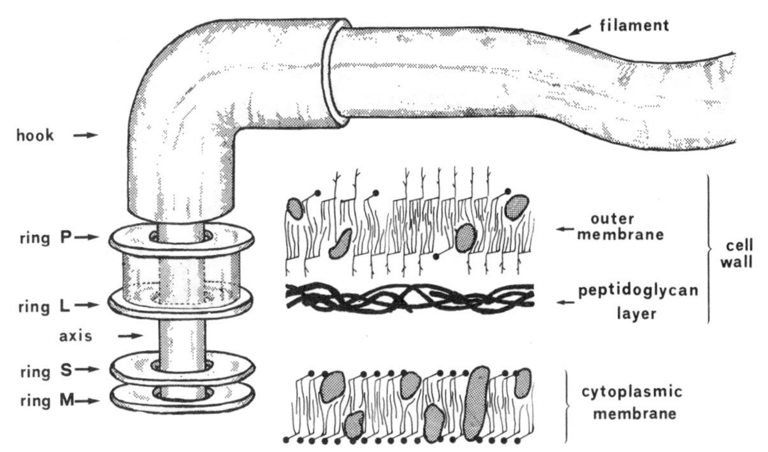

Fig. 2.3 Schematic model of flagellar basal body in gram-negative bacteria (*E. coli*), showing its relationships with the cell envelope structures. (From De Pamphilis and Adler [35, 36].)

having a length of 1.1 μm [72]. The transition from the normal superhelical form (2.3 μm) to the curly form (1.1 μm) has been observed to occur also under normal conditions during bacterial movement, and, as will be explained later, it plays a precise role in the swimming behaviour of a bacterium.

The hook is a structure located at the base of the filament and normally has an angled or curved shape. It has a slightly larger diameter than the filament and is about 900 nm long. Its function is not very well understood, but it probably acts as a *universal joint* allowing a more efficient transmission of the movement from the basal body to the filament and stabilizing the filament structure against environmental changes. A rod arising from the proximal end of the hook passes through the bacterial envelopes and connects with basal structures.

The basal body is a complex structure responsible not only for the attachment of the flagellum to the bacterial cell but also for the generation of its movement. Its ultrastructure (Fig. 2.3) reveals that, in a gram negative bacteria (i.e., bacteria possessing a cell wall with an internal peptidoglycan layer and an external lipolysaccaride membrane), it has four rings almost identical in diameter. The innermost or M (membrane) ring is attached to the cytoplasmic membrane and is the origin to the rod of the flagellum. The S (supermembrane) ring is visible just above the cytoplasmic membrane and is apparently unconnected to any other cellular structure. The two additional rings, P (peptidoglycan) and L (lipopolysaccaride), are embedded in the corresponding layers of the cell-wall and provide anchoring sites for the rod in its passage through the bacterial wall. The structure of the basal body in gram positive bacteria (e.g., *B.subtilis*) is similar but more simple than in the gram negative varieties. It shows only one pair of rings which, positioned at the base of the rod, seem to be analogous to the S and M rings of *E.coli* and are therefore referred to with the same letters

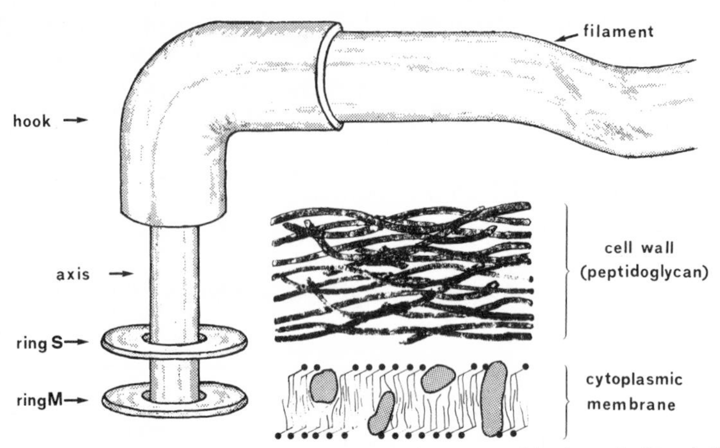

Fig. 2.4 Schematic model of flagellar basal body in gram-positive bacteria (*B. subtilis*), showing the relationship with cell membrane and cell wall. (From De Pamphilis and Adler [35, 36].)

(Fig. 2.4). The M ring of *B.subtilis* is located in the cytoplasmic membrane and the S ring is attached to the inner part of the peptidoglycan layer, where it is probably in contact with teichoic acids (components of the single-layered cell wall of gram positive bacteria). No additional rings connect the rod to the wall.

The study of the ultrastructure of flagella has therefore demonstrated a basic structure common to gram positive and gram negative bacteria, and the presence in both groups of the S and M rings has suggested that one or both of them could be involved in generating movement. Recent results have allowed a better understanding of the relationship between flagellar structure and movement and seem to indicate that the M ring is the component responsible for generating movement.

2.2 Molecular structure of the bacterial flagellum

The technique for isolation of intact flagella has not only allowed morphological study but also close chemical examination of their constitution, which is necessary for a better understanding of the complexity, the function and the regulation of the organelles.

From a chemical point of view, the filament is made up of identical subunits of a single protein called flagellin that can be easily separated from the hook–basal structure of intact flagella by low pH or by increasing the temperature to 60°C [3]. The flagellin subunits are able to reassemble *in vitro* to form filaments very similar to the one seen *in vivo*. The flagellin chemical constitution is characteristic of a given bacterial strain and antibodies induced against one bacterial strain will not react with flagellin from other strains, thus helping to classify bacteria on the basis of the antigenicity of their flagella. In order to understand the relationship between the molecular constitution, the shape, and the antigenicity of the flagellum, amino acid sequences of flagellin have been determined in a few bacteria [31, 76], but so far these studies have not been able to clarify the problem sufficiently. The molecular weight of the flagellin subunits, as expected, varies from one bacterium to another, being 54 000 in *E.coli* and 40 000 in some other species. Optical diffraction, seen with the electron microscope, has shown that a flagellum is a hollow cylinder and its wall (in a cross-section) appears to be made up of 11 longitudinal rows of flagellin subunits [78]. The angle between each subunit and the row axis is responsible for the helical structure of the flagellum, and this angle changes according to the different types of helical structure found.

The study of the biochemistry of the hook and basal body was initially complicated by the fact that these structures constituted only about 2 per cent of the intact flagellum, making it difficult to obtain a sufficient amount of material to work with. However, two types of mutant strains were found both in *E.coli* and *Salmonella*, one able to produce intact basal structures with no filaments [178], the other able to produce abnormally high amounts of the hook protein and form extended poly-hook structures. In the first case basal bodies could be obtained from the isolated membrane simply by detergent lysis, and in the second case poly-hook could be readily removed and purified from cells. When this material was subjected to polyacrylamide gel electrophoresis in the presence of sodium dodecyl sulphate (SDS), it could be separated into different polypeptides which could than be examined individually. Using this procedure the hook appears to be composed of a single polypeptide subunit with a molecular weight of 42 000 in *E.coli* and *Salmonella* and of 30 000 in *B.subtilis.*

The basal structure, as can be expected from the ultrastructural results, is chemically far more complicated. In *E.coli* it possesses at least 9 different polypeptides with molecular weights ranging from 9000 to 60 000 and possibly 3 other more labile bound components may be lost during the purification procedures [65]. Basal structures of gram positive bacteria have not been studied so extensively, but probably are of a less complex constitution (Table 2.1).

2.3 Regulation of synthesis and assembly of the bacterial flagellum

The structure and the assembly of a bacterial flagellum is genetically

Table 2.1 Flagellar proteins in *E.coli* (from [135])

Components (mol. wt.)	*Location*	*Gene*
54 000 (flagellin)	filament	hag
42 000	hook	fla region I
60 000	basal structure	fla region I
39 000	basal structure	fla region I
31 000	basal structure	fla region I
27 000	basal structure	fla region I
20 000	basal structure	not determined
18 000	basal structure	not determined
13 000	basal structure	not determined
11 000	basal structure	not determined
9 000	basal structure	not determined

controlled by sets of genes which have been defined and studied, particularly in *E.coli* [65, 135] and *Salmonella* [see 71]. Additional genes are involved in the control of the flagellar activity and in orienting movement of the cell in response to physico-chemical stimuli.

In the next section we will describe how the flagellar structure is genetically controlled and which mechanisms regulate the synthesis and assembly of the flagellar components. The study of the steps involved in this process is not yet complete, but some good examples have already come to light and may help in understanding the problem.

2.3.1 The flagellar genes of E.coli

The genes involved in flagellar motility in *E.coli* (as well as in other organisms) can be divided into three groups, according to their function. The first one comprises *fla* and *hag* genes whose end-products are all related to the flagellar structure: cells that possess one or more defective *fla* genes (*fla*$^-$ mutants) cannot assemble a proper flagellum. The second group consists of the *mot* genes which are responsible for the rotatory movement of the flagellum. *Mot*$^-$ cells possess structurally intact filaments which are unable to move. The third group is made up of the *che* genes whose function is to orientate the tactic response of the cell to a large number of stimuli. *Che*$^-$ cells fail to respond chemotactically and are unable to perform orientated motility. All these genes (about 30) have been mapped in defined areas of the *E.coli* chromosome and most of the gene products have been identified (Fig. 2.5).

Even though in this section we are more interested in studying the functional organization of *fla* and *hag* genes, a general discussion about the process of gene mapping and product identification will be conducted for all three groups of flagellar genes. The function of the *mot* and *che* gene products will be discussed in the sections dealing with the function of bacterial flagella and the regulation of cell movement.

Although the location of the genes on the bacterial chromosome can be obtained with standard genetic techniques, more sophisticated methods of gene cloning have been used for the identification of the gene products. In particular, fragments of the *E.coli* genome carrying the flagellar genes were inserted either into a *lambda* phage or a colicinogenic factor and introduced into cells capable of expressing only the acquired genome. In this way it has been possible to compare two-dimensional gel electrophoresis of the polypeptides synthesized under the direction of the flagellar genes introduced in permissive cells with those present in purified flagellar structures, allowing the recognition of specific genes coding for specific structures. It has been shown that the *hag* gene codes for flagellin, the *fla K* codes for the hook subunits, and all of the other genes of the *fla* group code for proteins of the basal structure. Furthermore, the product of *fla I* is necessary for the expression of the other flagellar genes. Among the flagellar genes, six are located in the region I of the *E.coli* chromosome and map between *gal* and *tpr*. The remaining genes are in the regions II and III between *zwf* and *supD*, together with *che* and *mot* genes (Fig. 2.5).

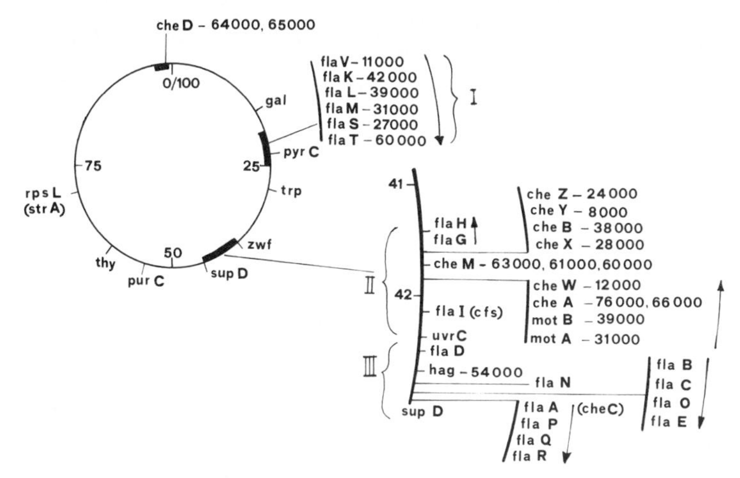

Fig. 2.5 The distribution of genes controlling flagellar activity on the genetic map of *E.coli*. Region I comprises the genes mapping between *gal* and *tpr*. (From *fla VK* and *fla T*.) The genes between *fla H* and *uvr C* are in region II and the remaining in region III. Alongside each gene is listed the apparent molecular weight of each gene product. Arrows indicate genes that are cotranscribed and the sense of transcription. (From [135].)

Little is known about the control of the genetic expression of flagellar genes. Some observations support the idea that the expression of one *fla* gene triggers the expression of the others. For instance, the product of the *fla I* gene is required for the expression of a number of operons of the *fla-hag*, *mot*, and *che* groups. However, other similar activators have not yet been identified [135].

The synthesis of the flagellar organelles is controlled by environmental factors, such as temperature, and is subject to catabolic control: it is repressed by glucose and a variety of catabolites but is activated by cAMP [7, 39]. This is not unexpected since in other metabolic bacterial systems cAMP acts as a non-specific gene activator overcoming the so-called 'catabolite repression' [110]. All the observations so far suggests that the effect of cAMP, as in other systems, is due to the formation of a complex between cAMP and its receptor protein (cAMP-CRP). This complex apparently binds to flagellar genes and activates their expression. The binding site is called CFS (constitutive flagellar synthesis). It maps close to the *fla I* gene and is likely to be part of it.

The assembly and growth of the filament of the bacterial flagellum is known in some detail. Its growth initiates from the distal part of the hook structure where flagellin subunits first assemble to form the hollow, wave-like cylinder. By using radioactive leucine and electron microscope autoradiography, it has been shown in *B.subtilis* that filaments are elongated by adding subunits at its distal end [44]. The model of filament growth which corresponds best to the experimental data implies that flagellin subunits are often synthesized in the cell and then transported, through the central cavity which runs inside the flagellum, in the growing point (Fig. 2.6). No flagellin pool has been

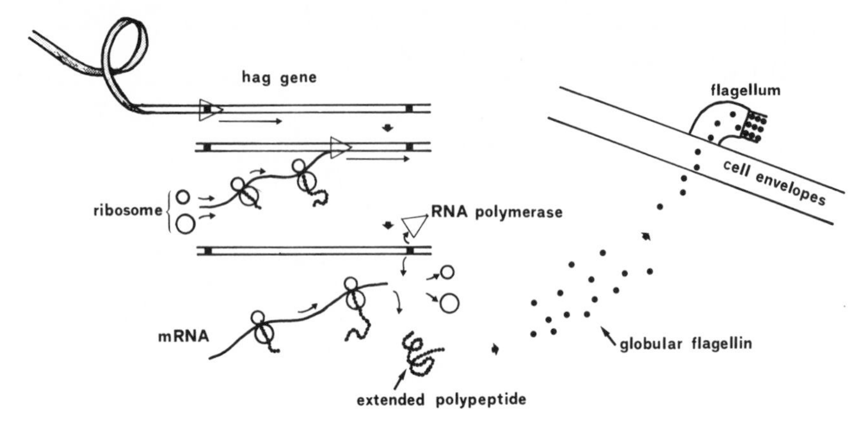

Fig. 2.6 Schematic model of flagellar elongation. The *hag* gene is transcribed in a specific mRNA that directs the synthesis of flagellin. Flagellin subunits reach the growing front of the flagellum through a central cavity and here they are assembled into the filament.

found inside the bacterial cytoplasm; therefore all the subunits must have been transported immediately after being translated.

2.4 The function of bacterial flagella

In this section we will describe how bacterial flagella move and how the motility force is generated and propagated to the filament.

When a flagellated bacterium is observed to move in a microscopic field, a rotatory movement of the body in the opposite direction of the fagellum is evident. Unfortunately, such a direct examination cannot give conclusive evidence to the type of movement. Until a few years ago it was commonly believed that movement was generated by waves originating from the base of the flagellum and propagating through it. However, a correct interpretation of existing evidence on the types of bacterial movement was made by Berg and Anderson in 1973 [20], who postulated that the movement of the bacterial flagella was of a rotatory type. This assumption was proved to be true by a clever experiment by Silverman and Simon [134] with poly-hook mutants of *E.coli* (i.e., bacterial unable of synthesizing the filament and which possess abnormally long hooks) cultivated in a medium in which they produce only one polyhook instead of six. Obviously, when these cells were completely free they were non-motile because of the absence of a filament. When antibodies against the hook protein were used to attach the hook alone to a glass slide (leaving the bacterial body free to move), the cells were seen to rotate at 10–20 revolutions per second. This experiment showed that he hook (to which normally the filament is attached) is rotating in the plane of the cell wall. This rotatory mechanism evidently is powerful enough to rotate the entire bacterial cell when the hook itself is immobilized. Similar results were obtained with straight mutants and were used as a direct proof that the flagellar movement is of a rotatory type. Rotation of attached bacteria could be clockwise or counter-clockwise, suggesting that the modulation of the direction of rotation could be the basis for the orietation of the movement of the cell.

It is easy to conceive how the rotation of a single flagellum (acting as a propeller) can move a cell, but it is more difficult to understand what happens in peritrichous bacteria where a large number of flagella are present, each originating from a different point of the bacterial wall. Microscopic examination has shown that swimming peritrichous bacteria possess a single bundle of flagella flowing backwards along the long axis of the cell body. Each flagellum in the bundle rotates synchronously with, yet independently of, the other flagellum. Under normal conditions, there are no connections or bridges between flagella. In fact, when adjacent flagella from normally motile *E.coli* are joined by means of bivalent antibodies against flagellin, cell movement completely stops.

At this point, considering the ultrastructural appearance and the motility behaviour of the bacterial flagellum, a dynamic model can be imagined as follows: the M ring, which is inserted in the lipid layer of the cell membrane, is able to rotate. Since it is connected to the rod, the M ring rotation will be transmitted to the rod and thereby propagate outside the cell to the semi-rigid helical filament. The rotation of the filament acts as a propeller and makes the cell swim. The remaining structures, and particularly the P and L rings, are not directly connected with the rotatory motor but presumably ease the passage of the rod through the bacterial envelopes.

The last question to be answered refers to the source of energy necessary for feeding such a type of rotatory motor. As already shown by Larsen and co-workers [83], by using mutants of *E.coli* and *Salmonella typhimurium* which are blocked in various stages of oxidative phosphorylation as well as employing specific inhibitors, ATP is not the direct source of energy for bacterial motility. The energy seems more likely to be derived from one or more intermediates of oxidative phosphorylation. Recent work done with *Streptococci* [93] and *Rhodospirilium rubrum* [60] has substantially confirmed these results and shown that rotation of the M ring, and therefore of the flagellum, is dependent on proton (H^+) movement from outside to inside the cell across the cell membrane (i.e., from the proton pump). More work needs to be done to explain how the proton passage across the cell membrane is able to activate the flagellar motor.

2.5 Patterns of bacterial movement

Flagella must be regarded as very effective propellers for the bacterial cell, because in normal conditions they are capable of producing orientated cell movement of an incredible speed. For instance, *E.coli*, which is about 2 μm long, can cover a distance of at least 20 μm in one second (this can be likened to a horse running at 180 km per hour). Within a microscopic field an *E.coli*, free from attracting or repulsing stimuli, can be observed to move in straight lines with abrupt changes in direction (tumbling), resulting in a three-dimensional random walk. It is

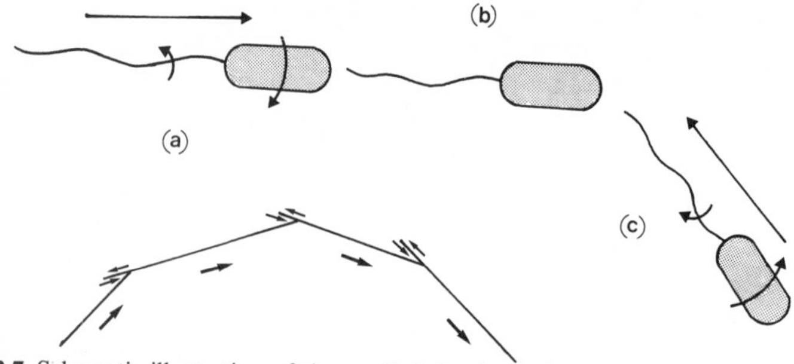

Fig. 2.7 Schematic illustration of the motile behaviour of a polar flagellated bacterium. The bacterium swims in a forward direction by rotating its flagellum (a). In (b) the flagellar motor stops and swimming is arrested. In (c) the reversal of rotation causes a backward motion. The black line represents the trace of a random motion of a polar flagellated bacterium moving forward and backward alternatively.

important to stress that without 'tumbling' a bacterium cannot change direction.

Observation of monotrichous and peritrichous bacteria shows that the mechanism of directional change always requires the reversal of flagella rotation, but there are differences in their respective behaviour patterns. The swimming behaviour of a polar flagellated bacterium such as *Pseudomonas citronellolis*, studied by Taylor and Koshland [155], consists of a run in a forward direction in which the bacterium is pushed by the propulsive force of the flagellum, followed by a reversal of the direction of rotation by which the bacterium is pulled back. The same process is repeated many times, and since change in cell orientation sometimes is as much as 180°, but frequently of a smaller angle, the result is a random walk motion (Fig. 2.7). During the forward run the flagellum rotates in a counter-clockwise direction; the opposite occurs during the backward run.

In free-swimming multiflagellated bacteria, however, a typical *backing up* process does not occur. When a *S. typhimurium* or *E.coli* cell swims in a given direction, as we have already seen, all the flagellar filaments are held in a compact bundle. The bundle is disrupted when the reversal of the rotation occurs and the flagella fly apart for a brief period of time, during which a random reorientation of the cell axis occurs. This period of reorientation corresponds to a tumble. Rotation movement is then continued and the cell can swim again, with its flagella in a bundle, but in a new direction (Fig. 2.8). When this tumbling phenomenon occurs, it results in a randomly orientated motility, in the absence of any chemotactic signals [91].

During the reversal of rotation there is not only a change in the direction of the movement, but also in the helical configuration of the filament from a superhelical to a curly form. This implies a change in the packing of the flagellin subunits so that the structure flips from a left-handed to a right-handed helix. This is likely to be due to the torsional stress occurring during the reversal of rotation, and the transition propagates out from the base of the flagellum where the torsion is

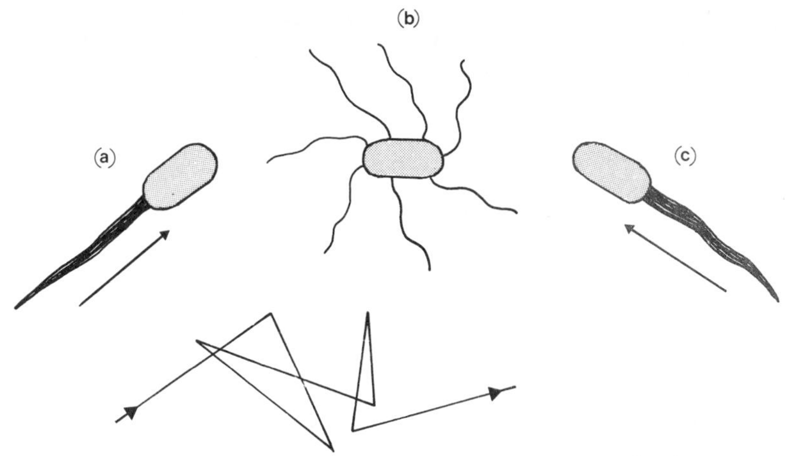

Fig. 2.8 Schematic illustration of the motile behaviour of a multiflagellated bacterium. The bacterium swims in a forward direction with flagella in a bundle (a). When the rotation is reversed, flagella fly apart (b) before resuming a new bundle disposition (c). The black line is the trace of a random motion of a multiflagellated bacterium.

greater [92]. How these changes are directly involved in motility is at the moment a matter of speculation.

2.6 Regulation of bacterial movement

When flagellated bacteria are allowed to move in a liquid or semi-solid medium such as an agar plate, where gradients occur in the concentrations of attractant or repulsive substances, they show an orientated motility and are able to move towards the area where there is a greater concentration of attractants or a lower concentration of repellants. This phenomenon is called the tactic response and is exhibited by wild type strains of flagellated bacteria in response to chemical stimuli (chemotaxis), changes in temperature (thermotaxis), and modification of light intensity (phototaxis) [5, 162]. The evolutionary advantages of a system that allows the orientation of movement towards a food source or away from a toxic environment is evident and, in general, can make a flagellated bacteria more competitive with respect in non-flagellated microorganisms.

The movement of a bacterium subjected to chemotactic stimulation can be studied using a microscope equipped with a special device adapted by Howard Berg [19], which is able to follow and record the movements of a single cell in a three-dimensional space. The tracks obtained from a bacterium moving towards an attractant show that the movement consists of directional runs and random tumblings similar to those present in non-orientated motility. However, the frequency of tumbling is lower and the time of single runs is increased. The result of this behaviour is a net movement towards the chemotactic stimulus. The opposite occurs when a bacterium is escaping from a repellant: in this case the tumbling frequency is higher and straight runs are shorter. A similar pattern of motile behaviour is present in polar flagellated micro-organisms, except that tumbles are substituted by *backing up*. From these experiments it is evident that orientated motility is obtained through the control of the tumbling activity and the length of the runs towards, or away from, a tactic stimulus.

Bacteria are able to sense and respond not only to spatial gradients but also to temporal gradients of active substances; they respond with a modification of their movement when different concentrations of repellant or attractant are mixed very rapidly to the medium in which they live [90]. The excited state, i.e., the increased tumbling frequency, is transient and, after a set period of time, bacteria return to the unstimulated state. This implies that they respond only to the differences in the temporal concentration of added chemicals but not to the absolute amount of such chemicals. Furthermore, they show a kind of short-term memory. When bacteria have been stimulated by a certain concentration of attractant and then returned to a non-stimulatory concentration, they will not react to the stimulatory concentration if challenged within a short period of time after the original stimulation. This indicates that they are able to sense and remember, for a short period, the level of the stimulant (see [5, 47]).

Having defined the general characteristics of chemotactic regulation of bacterial movement, we will now describe the components of the system and try to define how they can be correlated with the oriented movement of the flagella. Many of the results to be discussed have been obtained using mutants defective in some aspects of chemotaxis, generally called *che*$^-$ mutants. The *che* system in *E.coli* is made up of a complex of genes mapping in three different regions of the chromosome, all involved in some steps of chemotactic response. For most of them, using the techniques described for the *fla* system, the gene products have been identified as polypeptides with molecular weights varying from 8000 to 76 000 and their functions have been recognized.

In a very simple scheme, the chemotactic system of bacteria is formed by: receptors able to recognize input stimuli, a machinery for the processing of signals coming from the receptors, and a response regulator able to modulate the movement of the flagella (Fig. 2.9).

The receptors are the part of the chemotactic system most understood both in terms of the number of different chemosensors and in terms of their chemical nature and function. They are able to detect the attractant itself, i.e., the attractant does not need to be modified to be detected by the receptors. They differ from other receptors of the bacterial cell (i.e., receptors for bacteriophages, bacteriocin, etc.) by the former's ability to inform the flagella that a stimulant has been detected. In *E.coli* more than twenty different types of chemoreceptors have been identified; some of them are responsible for positive chemotaxis (toward an attractant), and others for negative chemotaxis (away from a repellant). In the first group are included sensors for sugars and amino acids, and in the second those for fatty acids, alcohols, indoles, pH variation and metallic ions. Remarkably, most of the attractants are useful for cell

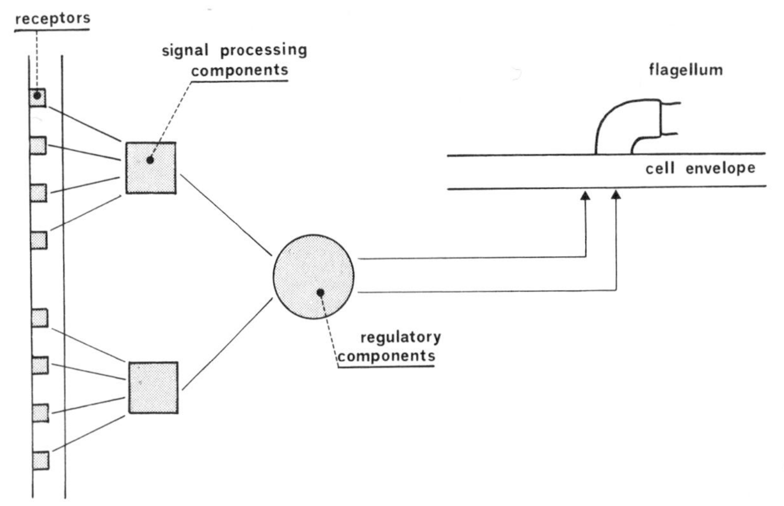

Fig. 2.9 Hypothetical schematic representation of the chemotactic system in bacteria. Receptors, after the binding of a stimulatory molecule, transmit the signal to processing components which transform it into appropriate changes in flagellar rotation by means of regulatory components.

metabolism and most of the repellants are dangerous. In at least one case (glucose), there is not a unique specificity between the attractant and the receptor, since this sugar can also be sensed by galactose and mannose receptors.

2.6.1 The nature of the chemoreceptors

Chemoreceptors are proteins which, in order to be able to interact with an external stimulus, must be located on the outside of the cell membrane, at least for some of the time. In addition to their chemoreceptor functions, some of them also act as *binding proteins* necessary for the transportation of specific molecules across the cell membrane. This is the case for the galactose-binding protein of *E.coli* that operates in the α-methylgalactoside transport system and is, at the same time, the chemoreceptor for D-galactose, D-glucose and D-fucose [83]. Another example is the ribose-binding protein of *S.typhimurium*, which is also the chemoreceptor for the same sugar [15]. Binding proteins can easily be purified from the bacterium by osmotic shock, because they are present in the periplasmic space (between the cytoplasmic membrane and the cell wall) loosely bound to the cell membrane. Their interaction with the sugars they bind can therefore be studied in *in vitro* systems.

In order to function as chemoreceptors, the *binding proteins* must be able to interact with components of the signal-processing system which transmits the signal from the receptors to the flagella (Fig. 2.10). It has been postulated that a conformational change in the binding protein following contact with stimulants is responsible for the stimulation of the signal-processing system. At least in the case of the galactose-binding protein, it has been shown that the formation of the complex

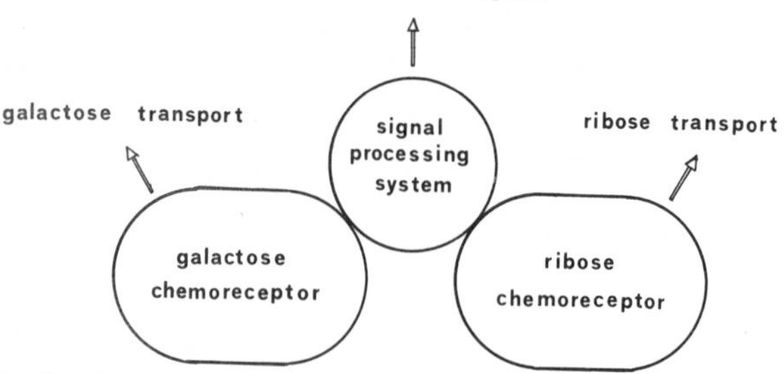

Fig. 2.10 The function of galactose and ribose receptors in *E.coli.* After binding with the sugar, receptors both stimulate the signal processing system and act in transporting the ligand into the cell.

with its substrate induces a conformational change in the protein structure consistent with this hypothesis [25].

An additional receptor for sugars has been identified in *E.coli* as enzyme II of the glucose phosphotransferase system, an enzyme which is tightly bound to the cell membrane and catalyses the phosphorylation of sugars by a phosphorylated protein [8]. More recently the membrane-bound Mg^{2+}, Ca^{2+}-dependent adenosine triphosphatase (Mg^{2+}, Ca^{2+}-ATPase), which is involved in the systhesis of ATP and in the final steps of oxidative phosphorylation, as well as in other metabolic pathways, has been shown to be the receptor for divalent cations in both *E.coli* and *S.typhimurium* [179].

2.6.2 The signal-processing system

The nature of the system for processing signals from the chemosensors and transmitting them to the flagella is almost entirely unknown. However, an insight into some of the biochemical events involved comes from the discovery by Adler and Dahl [6] that *E.coli* requires the amino acid L-methionine for chemotaxis. In *E.coli* autotrophs for methionine (i.e., mutants unable to synthesize methionine by themselves), the amino acid must always be present in the medium, otherwise they lose their ability to tumble and do not respond either to attractants or repellants. Methionine is used in the form of S-adenosylmethionine which is a well-known methyl donor to proteins [16]. Recent work by Springer, Goy and Adler [143] has given evidence that methylation of cytoplasmic proteins, called methyl-accepting chemotactic proteins (MCPs), is essential for the processing of information coming from the chemosensors. These MCPs constitute at least two functional units, each of which is necessary to process data from two different groups of chemosensors (identified by their sensitivity to different molecules). Furthermore, it has been shown that the products of the *che D* and *che M* genes are methylated in response to chemotactic stimuli and must therefore be regarded as MCPs. The methylation of the *che D* and the *che M* gene products requires the expression of two other genes of the *che* group: *che X* and *che W* [136].

After being processed, the input signal may be directly transmitted to the flagella to modify the speed and the direction of rotation of the rotatory motor. It could also interact with the flagellum via other proteins which are located in the cell membrane and control the rotative motor of the cell. These proteins have been identified as the end-products of the *mot A* and *mot B* genes; they are integral proteins of the cytoplasmic membrane and are probably located around the base of the flagellum [137].

Topics for further reading

The following references pertain mainly to general aspects of flagellar motility in bacteria:

Berg, H.C. (1975), How Bacteria Swim, *Sci. Amer.*, **233**, 36–44.

Silverman, M. and Simon, M.I. (1977), Bacterial flagella, *Ann. Rev. Microbiol.*, **31**, 397–419.

Simon, M., Silverman, M., Matsumura, P., Ridgway, H., Komeda, Y. and Hilmen, M. (1978), Structure and function of bacterial flagella, in *Relations between Structure and Function in the Prokaryotic Cell* (ed. R.Y. Stanier, H.J. Rogers and J.B. Ward), Cambridge University Press, Cambridge.

MacNab, R.M. (1978), Bacterial motility and chemotaxis. Molecular biology of a behavioural system, *CRC Crit. Rev. Biochem.*, **5**, 291–342.

The genetics of motility is effectively treated in:

Iino, T. (1977), Genetics of structure and function of bacterial flagella, *Ann. Rev. Genet.*, **11**, 161–82.

Parkinson, J.S. (1977), Behavioral genetics in bacteria, *Ann. Rev. Genet.*, **11**, 397–414.

For chemotaxis the following reviews may be consulted:

Adler, J. (1975), Chemotaxis in bacteria, *Ann. Rev. Biochem.*, **44**, 341–56.

Chet, I. and Mitchell, R. (1976), Ecological aspects of microbial chemotactic behaviour, *Ann. Rev. Microbiol*, **30**, 221–39.

Frere, J.M. (1977), La chimiotaxie chez les bactéries, *Bull. Inst. Pasteur*, **75**, 187–202,

which, apart from its scientific interest, can be a useful exercise in the French language.

Additional articles on different aspects of motility in bacteria are contained in Section 1 (pp. 29–113) of the book Cell Motility *(ed. R. Goldman, T. Pollard and J. Rosenbaum), Cold Spring Harbor Laboratory, Cold Spring Harbor (1976), which refers to primitive motile systems.*

3 The motility system of eukaryotic cells

The system devoted to the generation of motility in eukaryotic cells is complex but made up of remarkably similar elements in all the cells studied so far. The components of the motility systems are evidently able to adapt themselves to as many different situations as can be found, e.g., in a free-living amoeba, in a mammalian cell or in a cell of a flowering plant. This similarity implies that the major components of this system are coded by genes that have been conserved through the various steps of the cell's evolution.

The structures able to generate movement in eukaryotic cells may be classified into two main systems according to their morphological characteristics and their protein composition. The first system is formed by microtubules whose major protein component is tubulin and the second by microfilaments formed mainly by a protein called actin. Each system is responsible for different types of movement. Microtubules control ciliary and flagellar movements, migration of pigment granules in chromatophores, bending of axostiles, etc. Microfilaments control amoeboid movement, cytoplasmic streaming, cytokinesis, etc. Other types of movement, such as chromosome separation during cell division, are at the moment not yet definitively attributed to either microtubules or microfilaments.

Many other proteins besides the major components have been identified as being part of the motility system. Some of them, such as dynein and myosin, interact directly with tubulin and actin respectively in generating movement. Many others are devoted to regulatory functions and to making connections between microtubules, microfilaments and other cellular structures. Although from a morphological and biochemal point of view the two systems are distinct, there is evidence that from a functional point of view there may be overlapping and cooperation. For instance, the microtubular network inside the cell may serve as a frame for the attachment of microfilaments or, although less likely, there may be cooperation between the two systems to produce a contractile force.

Having roughly defined the components of the motility system, in the next chapter we are going to discuss in greater detail their characteristics, their interactions and how they function.

3.1 Microtubules

The recognition of microtubules as distinct structures inside the cell and particularly inside the flagellum has been made possible by the use of the electron microscope. In 1946, Jakus and Hall [75] demonstrated the presence of tubular structures uniform in diameter in *Paramecium* cilia. Subsequently, tubular structures have been found in almost all cells studied so far. They are recognized as fundamental structures characteristic of the eukaryotic organization. However, the presence of microtubular structures in the cytoplasm of prokaryotic spirochetes has also been reported [69]. These microtubules are morphologically and chemically similar to the eukaryotic ones [94]. This discovery lends support to the '*exogenous hypothesis*' for the origin of eukaryotic microtubules, which explains the presence of cilia and flagella in higher cells as having been acquired through the symbiotic association between spirochetes containing tubules and non-flagellated cells.

Cellular microtubules are long cylinders about 24 nm in diameter with a central cavity 15 nm wide. In the majority of cells they are only a few μm long, but in some specialized cells, such as the motor neurons of the central nervous system, they can be as long as several centimetres. The microtubule wall (about 5 nm in thickness) is made up of longitudinal protofilaments of globular subunits (clearly evident under the electron microscope after negative staining) chemically composed of a protein called tubulin (Fig. 3.1). Apart from the direct generation of cell motility, they are involved in other cellular processes more or less correlated with it, such as the maintenance of cell shape, the transport of intracellular materials, the secretion of cellular products, the movement of chromosomes at cell division, and perhaps in sensory transduction and the movement of cell membrane components [107, 149]. Microtubules can be found as scattered elements within the cell

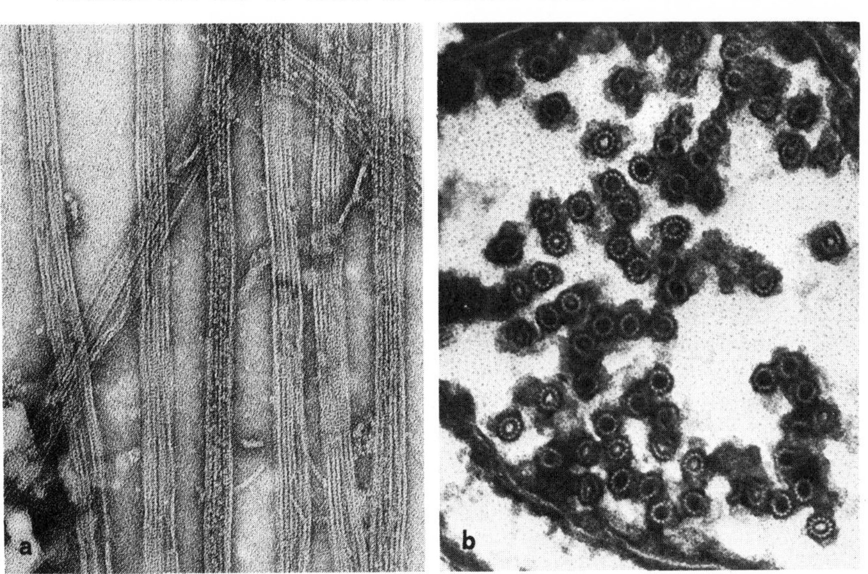

Fig. 3.1 Electron micrograph of microtubules from marginal band of newt erythrocytes. In (a) the organization of microtubule subunits into rows is evident in a longitudinal section; in (b) the 13 subunits forming the wall of each tubule are seen in a transverse section. Negative staining, 200 000 ×. (Courtesy of Dr. G. Monaco.)

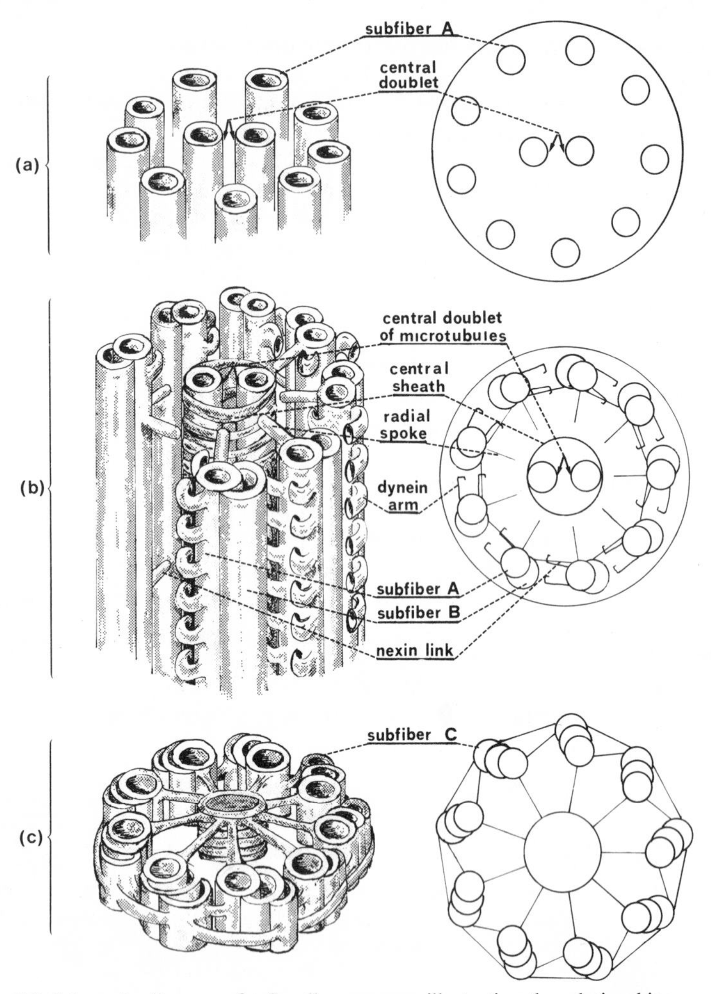

Fig. 3.2 Schematic diagram of a flagellar axoneme illustrating the relationships between the components at the flagellar tip (a), in the middle of the flagellum (b), and at the basal body (c).

cytoplasm or they can be seen as more organized structures.

In order to understand microtubular-dependent cell motility, we will discuss the relevant aspects of microtubular organization in eukaryotic cilia or flagella, in the contractile axostyle of certain flagellated cells, and in the microtubular network in the cytoplasm of animal cells.

3.1.1 Microtubule organization in cilia and flagella

Eukaryotic cilia and flagella are specialized structures projecting from the cell surface, which are capable of movement. The two types of organelles are structurally identical, showing minor differences in their function. They can be easily isolated from the cell in large quantities, sometimes simply by vigorously shaking the cell suspension. It has therefore been relatively easy to study their structure and chemical composition.

The structural patterns show similar features in almost all cells studied so far, that is, a ring of microtubules (usually doublets) surrounding a pair of central tubules (see [139]). The tubular structures are interconnected by 'bridges' and together they constitute what is called the flagellar axoneme. The characteristics of the tubules and bridges of flagellar axonemes are now known in some detail and can be seen in Fig. 3.2. The centre is occupied by a pair of tubules enveloped by a sheath formed by rings of protein at an oblique angle. The walls of the central microtubules are composed of thirteen longitudinal rows of subunits. Around the central structure there is a circle of nine double tubules. In each doublet one tubule (subfibre A) has a complete wall (formed by thirteen protofilaments, Fig. 3.3), and the other (subfibre B) is sickle-shaped and attached to subfibre A. Its wall is formed by a smaller number of protofilaments, about 10. Each subfibre A is connected to the B subfibre (of its neighbouring doublet) by links of a protein called nexin. *Radial spokes* made of protein protrude from the central-facing surface of the subfibre A towards the central structure. In addition, from each complete tubule originate two double rows of lateral arms, made of a protein called dynein, about 14 nm long and directed towards the neighbouring doublet. These projections are regularly spaced along the longitudinal axis of the A subfibre and, as we shall see later, are very important for the generation of movement.

The pattern we have described above refers to the axonemal structure that can be seen in the central part of the flagellum (i.e., excluding the two ends). At the base and tip of the flagellum, this pattern is somewhat different. At the tip, the central pair of tubules may be longer than the surrounding tubules (narrowing the end of the flagellum). Frequently the B subfibres terminate before the A subfibres so that the peripheral doublets become single near the tip (Fig. 3.2). The projections from the A subfibres disappear as they become single [129].

A specialized structure known as the *basal body*, from which the flagellum originates, is present in the cell cytoplasm at the base of the flagellum. This structure may vary in different cell types but common

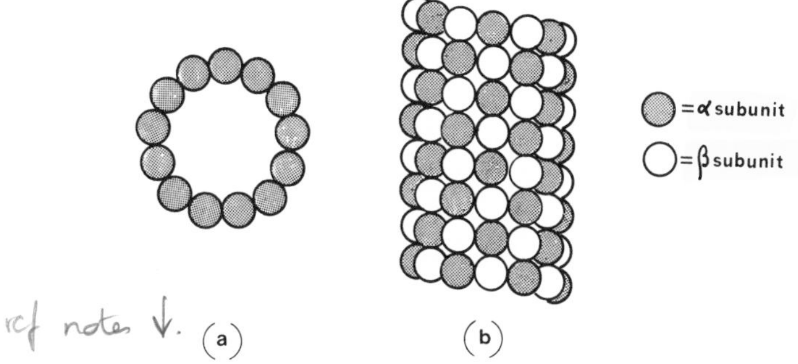

Fig. 3.3 Schematic representation of a cross-section (a) and a side view (b) of a microtubule showing the tubulin subunit arrangement.

features can be found. In the basal body of the ciliate protozoon *Tetrahymena piriformis*, in addition to subfibres A and B a new fibre C of about 10 protofilaments is present, thus forming a peripheral triplet. These triplets are linked together by means of lateral projections and to the central ring through radial projections (Fig. 3.2). The flagellar basal body is morphologically similar to the *centriole* of animal cells. Both function as organizers of microtubular assembly, the first for cilia and flagella and the second for the microtubules of the spindle.

Another interesting part of the flagellum is the *collar* that is the point at which a flagellum emerges from the cell surface. Electron micrographs of transverse sections of the collar show that there are fibrous connections between the peripheral doublets and the internal layers of the cell membrane. These connections give a rigid anchoring point for the tubules to the cell membrane and are involved in flagellar movement.

Motility in cilia and flagella is due to the intrinsic properties of the axonema, and not to other components of the organelles (i.e., cell membrane, cytoplasm, etc.), as has been clearly demonstrated by Gibbons and Gibbons [55]. They were able to obtain intact axonemes from flagella of sea urchin sperm by membrane lysis using the non-ionic detergent Triton X-100. These isolated axonemes are able to move after the addition of ATP as the energy source.

3.1.2 Microtubules of the contractile axostyle

The contractile axostyle of certain flagellated micro-organisms is a further example of a motile structure whose movement is due to microtubules. The axostyle is a rod- or ribbon-shaped organelle that runs through the cytoplasm from the one end of the cell to the other (Fig. 3.4). It was originally described by Grasse in studies on symbiotic flagellates of the gut of termites and wood-feeding roaches [61]. The

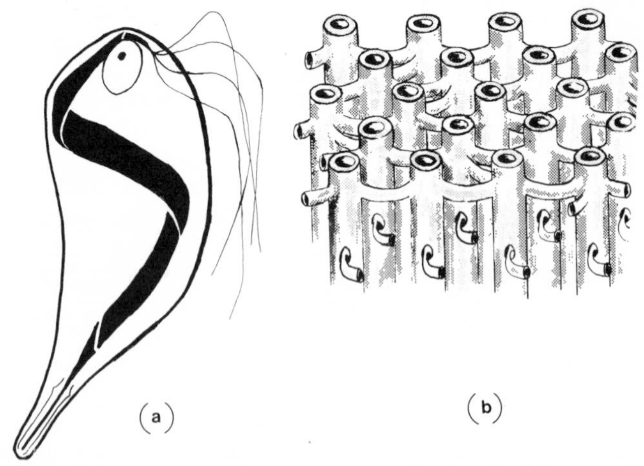

Fig. 3.4 The contractile axostyle of a *Metamonadida* flagellate. In (a) the relationships of the axostyle with the cell body are shown; (b) is a schematic representation of the arrangement of microtubules into parallel sheets as seen in a transverse section of the axostyle.

axostyle is capable of undulatory movement, and waves, originating from its anterior end and propagating backwards, flow through it. As a result of these waves, the shape of the cell is constantly changing and the movement produced by this change is responsible for cell movement [62]. The axostyles can be isolated from cellular contaminants by cell lysis using Triton X-100 and differential centrifugation, and after this treatment, like axonemes, they are capable of movement upon addition of ATP [99].

From an ultrastructural point of view, the axostyle consists of parallel sheets made up of a large number of tubules. The number of sheets and the number of tubules in each varies according to the species of the micro-organism. Smaller axostyles may have only two or three sheets with twenty to thirty tubules in each, while in large ones there are twenty to thirty sheets, each with about 150 tubules. This means that the total number of tubules may vary from less than 100 to more than 5000. Within a single sheet each tubule is separated from its neighbour by about 40 nm (centre-to-centre) and the sheets are approximately 30 nm apart [99]. Protein bridges, regularly spaced along the length of the tubules, connect the tubules within each sheet and also, probably, to other sheets. In addition, dynein arms form connections with microtubules of different sheets (Fig. 3.3). In this structure, motility is achieved because each sheet of tubules is free to slide with respect to adjacent sheets.

The axonemes of heliozoan axopods [161], the cytopharyngeal basket of the ciliate *Nassula* [163], the tentacle axonemes of suctorian ciliates [164, 67], and the microtubules in sperm tails of the coccid insect *Planococcus* [125] are other examples of motile structures formed by microtubules connected by links and forming regular arrays. (For a more detailed description of the above examples, the reader should consult specific studies; see References.)

3.1.3 Cytoplasmic microtubules of animal cells

The enormous increase of scientific interest shown by the biomedical world in the study of microtubules and microfilaments over the last ten years has led to the rapid expansion of ultrastructural and biochemical research on microtubules in mammalian cells. This has been made possible by the use of two techniques which have dramatically increased the ability to see such delicate structures as cytoplasmic microtubules. The first technique is based upon the use of aldehyde fixatives (e.g., glutaraldehyde). This has allowed better preservation of cytoplasmic organelles [127], leading to the discovery, in the majority of the cells examined, of elements similar in ultrastructure to flagellar microtubules. Cytoplasmic microtubules have been shown to be the major constituents of the mitotic spindle and also to be present in the cell cytoplasm, though dispersed differently in each cell type examined. Unless tedious techniques of tridimensional reconstructions of serial images are used, the electron microscope cannot give the overall microtubular pattern within an interphase cell. This has been achieved by specifically staining cytoplasmic microtubules of cells attached to microscopic slides with fluorescent antibodies against tubulin, followed by examining these slides with an ultraviolet light microscope [51, 172]. Originally, tubulin obtained from sea urchin sperm flagella (*Strongylocentrotus purpuratus*) was used to produce antibodies that were cross-reacting with mammalian tubulin (this immunological cross-reaction shows a great conservation of the amino acid sequence of tubulin during evolution). More recently, antibodies against purified mammalian brain tubulin became available and were used to confirm and extend the microtubular patterns previously obtained.

As an example of '*in vivo*' microtubule structures, we will examine the pattern of microtubule organization in cultivated animal cells. In fibroblasts, interphase cells show an incredible number of thin fibres, constant in diameter, within the cytoplasm. Often the fibres extend radially from the nucleus and terminate near the cell surface, or they run in irregular bundles parallel to the cytoplasmic membrane, particularly in the pseudopodal extension of the cell (Fig. 3.5). When microtubular disrupters, such as the alkaloid colchicine or low temperature or pressure treatment, are used the microtubules disappear and the cell loses its normal shape. After these inhibitors are removed, microtubule structures reform and the normal cell shape is regained [171, 172]. When cells approach the moment of division, cytoplasmic fluorescence (seen after treatment with fluorescent antibodies against tubulin) condenses near the nuclear membrane at two opposing points corresponding to the centriolar regions, and in pro-metaphase cells, two fluorescent asters can

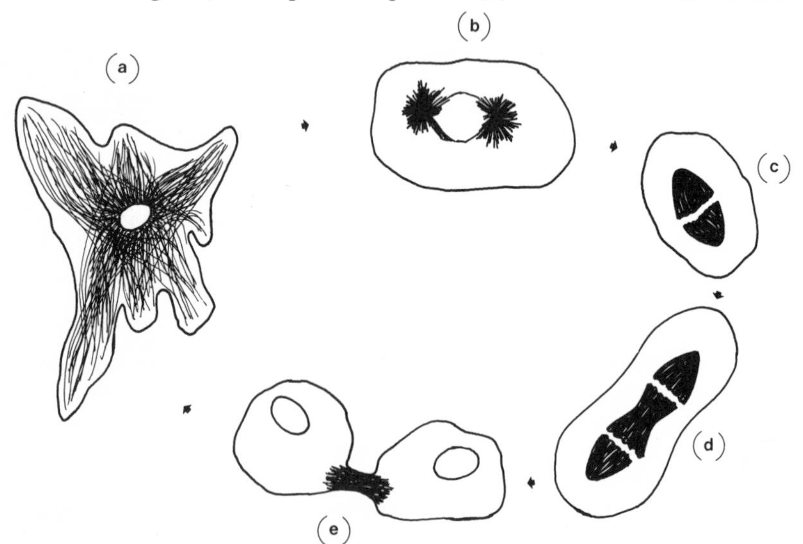

Fig. 3.5 Pattern of microtubule fluorescence during the cell cycle. In interphase (a) microtubules form a fine network inside cell cytoplasm. In prophase fluorescence condenses at the centriolar region (b). During metaphase (c) and anaphase (d), fluorescence is associated with spindle microtubules. At cell separation, only a fluorescent bridge between the two cells can be seen (e).

be seen. Further condensation occurs during the progression of mitotic events, and in metaphase cells an intense fluorescent spindle and a non-fluorescent equatorial plate of condensed chromosomes are seen. Spindle fluorescence starts to decline at the beginning of anaphase (due to the disassembly of the spindle) and, at the separation of the two daughter cells, only an intercellular microtubular bridge is seen. When the two daughter cells are completely separated, the usual pattern of thin cytoplasmic microtubules is restored [51, 171, 172]. These events show the existence of a cycle of cytoplasmic microtubules in mammalian cells parallel to the cell cycle itself, and indicate that microtubules can be assembled and disassembled in structures with different patterns according to the need of the cell.

Cytoplasmic microtubules are certainly not able to create cellular movement in the same way as flagellar or axostyle microtubules. However, it has been postulated that cytoplasmic microtubules may be involved with cell motility [59] because, when certain types of cultivated cells are treated with microtubular inhibitors, there is a general disruption in their orientated movement [22, 58]. In any case, it is fairly clear that in all amoeboid-type movement cytoplasmic microtubules play only a marginal role, if any at all.

3.2 Molecular components of microtubular structures

Microtubules are made up of a major protein called tubulin and other protein components which are generally known as microtubule–associated proteins.

3.2.1 Tubulin

Tubulin is an acidic protein with an isoelectric point in the range of pH 5.2–5.4 usually isolated as a 115 000 dalton dimer, also known as 6S tubulin. Treatment of tubulin dimers with denaturting agents such as urea, guanidine or sodium dodecyl sulphate (SDS) results in the formation of two different 3S subunits called α- and β-tubulin, similar in molecular weight (55 000), but different in amino acid composition [27] and in the sequence of the polypeptide chain [89]. As recently shown by Bryan and coworkers [28], the main difference between the two subunits is not due to post-translational modifications (although these may occur) since separate messenger RNAs, coding for each subunit, have been isolated from chick embryo brains.

However, post-translational modifications of the tubulin molecule do exist. One of these consists of the binding of guanine nucleotides (such as GTP) to the molecule, and two different binding sites with different affinities for GTP are present in each tubulin dimer [21, 150]. Another is a phosphorylation of serine residues due to a cAMP-stimulated protein kinase. This has been shown to occur in the β–subunits of cytoplasmic tubulin [43] and in the α–subunits of *Chlamydomonas* flagellar tubulin [114]. A third modification is the specific addition of a single unit of the amino acid tyrosine (or phenylalanine) to the carboxyl terminus of the α–subunit [18]. This reaction is due to a specific enzyme called tubulin-tyrosine ligase, which is widely distributed in avian and mammalian tissues [122]. The significance of the last two post-translational modifications is unclear. However, GTP addition is a general characteristic of tubulin and is related to microtubule assembly.

Apart from binding sites for nucleotides, phosphate, and amino acids, the tubulin molecule is able specifically to bind bivalent cations such as Mg^{2+} (at least 1.0 mole tightly bound Mg^{2+} per mole of tubulin dimer [108]) or again, with a defined stoichiometry, alkaloid drugs such as colchicine [175] and vinblastine [176]. This last property is particularly important since these drugs are able to inhibit microtubular activities selectively and have been largely used to explore microtubule-dependent reactions both *in vivo* and *in vitro*.

From the examination of tubulin from different cell types, it has become clear that most physico-chemical characteristics of this molecule have been maintained during evolution. All tubulins so far isolated have the same molecular weight and electrophoretic mobility. Furthermore, partial amino acid sequence data on α- and β-subunits of sea urchin sperm axonemal tubulin and chick embryo brain cytoplasmic tubulin show no difference in the first 25 residues of the N-terminal region for α-subunits, and a single amino acid change (at residue 7) between the two β–subunits [89]. Immunological characterization also shows similarities between tubulins from very different species and cross-reactivity has been demonstrated between mammalian tubulin and that of sea urchin [33, 51], the unicellular biflagellated algae *Chlamydomonas reinhartii* [115], and the cellular slime mould *Dictyostelium discoideum* [30]. Mammalian brain tubulin is also able to copolymerize (i.e., to form hybrid microtubules) with tubulin from yeast [170], moulds [132], and slime moulds [29]. These results demonstrate a substantial chemical similarity between (but not identity with) tubulins from different sources.

Another interesting question is the existence of different classes of tubulin within single cells, each of which may be necessary for building different structures. This implies the presence of different sets of tubulin genes, or different post-translational modifications of the same tubulin molecule. Evidence from different sources demonstrates the presence of partially different tubulins within a single cell. In *Naegleria gruberi,* an amoeboid protozoan which, under experimental conditions, can synchronously differentiate into a flagellar form, Fulton and coworkers [50, 81, 82] showed that flagellar tubulin is not antigenically related to the cytoplasmic one, and its synthesis is initiated *de novo* at the time of flagellar construction. Stephens [148] was able to separate A and B subfibres from outer doublets of sea urchin sperm flagella and showed that α- and β-subunits from each subfibre differ in their amino acid composition and in their tryptic peptide map. These results are consistent with the hypothesis of the existence of multitubulins within a single cell, each of them with special associative and structural properties needed for a particular microtubular function. If this is true, the tubulin system shows a remarkable similarity to that of another protein, actin, involved in motility. For actin, slight chemical variations exist not only among different biological species but also within single cells.

3.2.2 *Tubulin-associated proteins*

Tubulin-associated proteins are integral components of the microtubular system necessary for its construction and function. They may be involved in generating motility or in controlling the assembly and disassembly of microtubules, or in creating connections between microtubules and other cell components (see Table 3.1). Therefore, together with tubulin, they form a unitarian system both from a structural and functional point of view. Most of our present knowledge on associated proteins comes from studies on flagellar and ciliary axonemes and on brain microtubule organization.

The axoneme of cilia and flagella is a fairly complex structure composed of numerous proteins. Therefore, when complete axonemes of sea urchin sperm flagella are subjected to polyacrylamide gel electrophoresis in the presence of denaturing substances such as SDS, at least twenty to thirty different polypeptide bands can be identified. Two of them are more prominent: one is the well-known tubulin band (that under these conditions migrates as subunits with a molecular weight of about 55 000) and the other is a high molecular weight band (> 300 000), corresponding to a protein called dynein. Flagellar dynein, first purified by Gibbons [56], forms the arms attached to the A-subfibre in the peripheral doublets of the axonemes (see Fig. 3.2). It is an enzyme with an adenosine triphosphatase activity (ATPase). In sea urchin sperm flagella (as well as in other system), it appears to be composed of two isoenzymes that can be separated using techniques of differential solubilization in salt solution. Dynein 1 comprises the largest portion of the total amount of dynein and is the true constituent of the arms of the doublet tubules. The second isoenzyme, dynein 2, is present in small

Table 3.1 Some of the proteins known to interact with tubulin or microtubules

Components		*Activity*	*Source*
In the flagellar axoneme:			
Dynein 1		ATPase activity	sea urchin sperm flagella
Dynein 2		Generation of movement	,, ,, ,, ,,
Nexin		Forms linkages between adjacent doublets	,, ,, ,, ,,
In the cytoplasmic microtubules:			
High molecular weight proteins	MAP1 MAP2	Control of initiation and elongation of microtubules?	calf brain
tau		Promote assembly of tubulin molecules in rings and microtubules?	,, ,,
Protein kinase		Phosphorylation of tubulin subunits	,, ,,

amounts and does not seem to be directly involved in generating movement, but is involved in its control. This point, however, needs further clarification [57]. Much evidence indicates that dynein, through its ATPase activity, plays a major role in generating the active sliding movements between neighbouring doublet tubules which make flagellar movement possible.

Again using sea urchin sperm flagella, which are very suitable for procuring large amount of flagellar axonemes, another protein linked to the A subfibre has been characterized. This is an acid-soluble protein with a molecular weight of about 165 000, and it has been termed *nexin* [146] because it links adjacent doublets. Its role seems to be to limit the amount of sliding between two adjacent doublets during flagellar movement.

The understanding of the chemical composition and function of the other components of flagellar axonemes is by no means a simple task. Two different approaches have been used: the first consists of the purification of axonemal structure by physico-chemical methods, and the second is the analysis of the chemical composition and the ultrastructural appearance of mutant organisms in which the flagellar activity is impaired. It is worthwhile mentioning that although the second approach seems to be more reasonable and fruitful, it cannot be used easily with complex organisms. Linck [87], using complicated procedures involving differential solubility of axonemal components coupled with electronmicroscopical observation and electrophoretic analysis of the remaining structures, was able to obtain A-subfibres with the accessory structures attached. He also showed that ten minor protein components associated with the outer doublets are located in the A-subfibre. The study of *Chlamydomonas* mutants, in which flagella were paralysed or moved abnormally, has made it possible to correlate the absence of specific structures with specific functions proteins. For instance, by comparing electrophoretical patterns of wild type axonemes with those of a mutant with no radial spokes, the major constituent of the spokes has been identified as a protein with a molecular weight of 118 000. Mutants lacking central tubules and sheaths have permitted the identification of three polypeptides, with a molecular weight higher than 220 000, which compose the sheath [177].

It seems likely, however, that the complexity of axonemal components obtained as we have just described is largely underestimated due to the fact that the electrophoretic system used has insufficent resolution. In fact, using two-dimensional isoelectric focusing and SDS–gel electrophoresis, wild type axonemes of *Clamydomonas* have been resolved into at least 100 different polypeptides. Using the same technique on mutants lacking radial spokes, twelve proteins were discovered to have been missed [116]. These results show how complicated the system is and how many steps are needed before it can be clearly elucidated.

On the other hand, it is a pleasant surprise to find that the system of proteins associated with cytoplasmic microtubules of mammalian cells seems at the moment to be less complicated. This, however, may be due only to the fact that the microtubule–associated proteins (MAPs) that we know of are almost exclusively those of the nervous system cells. Since the discovery of simple methods for obtaining large amounts of tubulin from mammalian brain extracts (where it constitutes more than 20 per cent of the total protein content) through repeated cycles of assembly and disassembly at different temperatures [133, 174], brain tubulin has been the most studied among mammalian tubulins and most of the present knowledge of tubulin is based on it. In the assembly and disassembly purification procedure, other proteins copurify with tubulin in stoichiometric concentrations. These proteins have been called MAPs, and it has been shown that they are associated both with intact microtubules of the central nervous system and with cytoplasmic and spindle microtubules of other cell types.

Although there is considerable variation in MAPs obtained during tubulin purification in slightly different conditions in different laboratories, two sufficiently reliable groups of MAPs have been identified and characterized (see Table 3.1). The most prominent of these is made up of high molecular weight polypeptides which, in polyacrylamide gel electrophoresis, migrate around 300 000 daltons and have been called HMW [24]. These HMW MAPs (in some microtubule preparations) are about one-fifth of the tubulin content. Electron microscopic evidence shows that these MAPs are structural components of microtubules which form thin projections attached to the tubules at regular intervals along their length [101]. Due to their high molecular weight and their ability to form lateral arms, these proteins could be considered to be similar to dynein. However, no ATPase activity has ever been detected and there is no evidence that they are involved in microtubular movement. Another component of the MAPs is a group of proteins with a molecular weight around 70 000 daltons which in the tubulin preparation of certain laboratories appears in a single band and has been called *tau* [173]. Fluorescent antibodies against *tau* have shown it to be present in cytoplasmic microtubules and its action seems to be correlated with microtubule assembly.

A cyclic AMP-stimulated protein kinase, probably responsible for post-translational tubulin phosphorylation, can also be considered as an MAP since it is found associated with tubulin and maintains a constant stoichiometry with it during several cycles of purification through assembly and disassembly [140].

3.3 Microtubule assembly and its control

The assembly and disassembly of tubulin in microtubular structures is widespread in living cells. The control of this assembly and disassembly process is therefore of primary importance for the understanding of many cellular processes.

3.3.1 In vitro *assembly*

The study of *in vitro* assembly of neuronal tubulin has provided a useful source of information for understanding *in vivo* microtubular assembly and its control. In this section, we shall first examine the reaction of

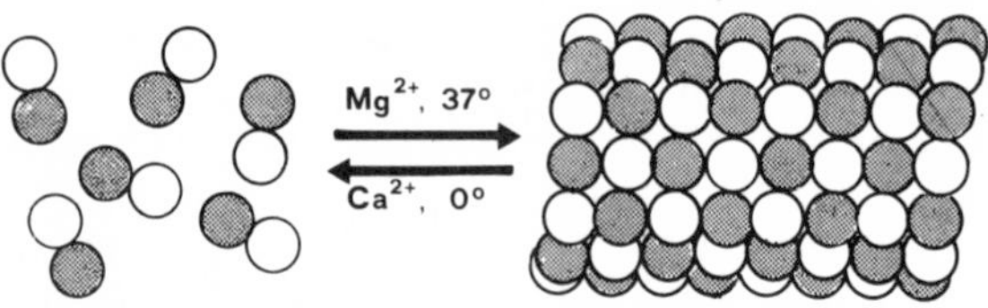

Fig. 3.6 The polymerization reaction of tubulin dimers into microtubules, showing the effect of divalent ions and temperature.

in vitro microtubule assembly and then what is likely to occur in *in vivo* systems.

Provided some basic conditions are fulfilled, neurotubulin dimers in solution, in the presence of associated proteins, can easily assemble. The assembly solution must have a moderate ionic strength, a slightly acid pH (around 6.6–6.7), nucleotides (GTP), and the divalent cations Mg^{2+}. The direction of the reaction is temperature-dependent: *in vitro* at physiological temperatures, tubulin is assembled into microtubules; these are rapidly disrupted when the temperature is lowered to 0°C (Fig. 3.6). The polymerization reaction can be followed by measuring light-scattering with a spectrophotometer, or viscosity modification of the solution with a viscosimeter. The assembled microtubules can be viewed by electron microscopy [53, 142]. As can be easily understood, this constitutes a simple *in vitro* system in which many parameters which may influence the polymerization reaction can be evaluated. Using this approach, it has been shown that amongst other things, high concentration of sucrose or glycerol increases the polymerization reaction and polyanions such as RNA or alkaloids, like colchicine, inhibit polymerization. By using the same methods, the requirement for ions and GTP has been carefully assessed. This was interesting because these substances may be regarded as potential controllers of *in vivo* tubulin assembly.

High concentrations of divalent cations inhibit microtubule assembly, and among them Ca^{2+} is particularly active as an inhibitor since quantities exceeding 1.0 mM cause a 50 per cent decrease in the reaction. GTP is necessary at a concentration at least equal to the molarity of dimer tubulin. It appears to be hydrolysed as tubulin assembles, and recently a GTPase activity has been found at the ends of growing microtubules [34]. GTP can be substituted by non-hydrolysable analogues (such as 5′–guanylylmethylene diphosphonate or GMPPCP), which are able to support tubulin polymerization, suggesting that the role of GTP in microtubule assembly is to provide a conformation of tubulin favourable for its polymerization. By cleavage of the terminal phosphate bond of GTP, the reaction may shift towards the formation of microtubules [17]. However, GTP is not used as an energy source because no source of energy is required for microtubular assembly, as demonstrated by the GMPPCP experiment.

These results were obtained with neuronal tubulin preparations in which microtubule-associated proteins were present. When microtubule formation is attempted with batches of purified subunits (i. e., without

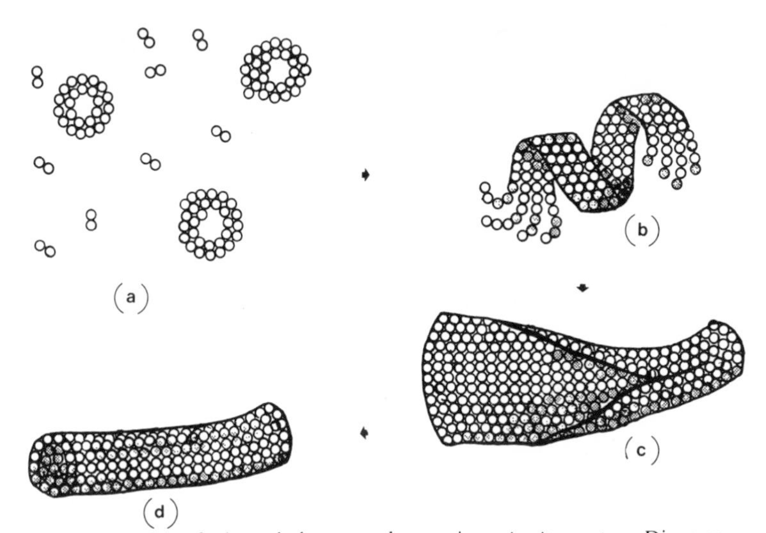

Fig. 3.7 Assembly of microtubules as can be seen in an *in vitro* system. Diagram shows: (a) = rings and spirals of tubulin together with free dimers; (b) = ribbon formation; (c) = folding of sheets into tubules; (d) = a completed microtubule. (From [112].)

MAPs), no polymerization occurs unless particular conditions are present (see [64, 66]). It has therefore been concluded that MAPs have a regulatory function in the reaction of tubulin polymerization, and many efforts have been made to clarify this function. Before dealing with the regulatory function of MAPs, we must briefly define how the polymerization reaction occurs. A model of microtubule formation is based upon the results of Kirschner's group [112], who have been able to follow the ultrastructural aspect of microtubule formation with a quantitative technique (Fig. 3.7). They showed that before the onset of polymerization a few rings and spirals of polymeric tubulin with a sedimentation coefficient of 36S are present, together with soluble dimeric tubulin. These rings have been shown to be necessary for the reaction because if they are removed by ultracentrifugation no polymerization can occur. When the reaction is started by adding GTP to the mixture and increasing the temperature to the required level, helical ribbons are formed within a few minutes. The tubulin dimers continue to attach themselves to the ribbons until a stage of thirteen protofilaments is reached. At this point, the ribbons fold up into a cylindrical shape to form a segment of microtubule [112]. Further elongation is caused by the attachment of dimers to the segment of microtubule (Fig. 3.8). Bearing in mind this process, the action of MAPs in microtubular assembly may be more clearly understood. It has been shown that *tau* is primarily involved in the formation of 36S polymeric tubulin (rings and spirals) without which microtubule formation is impossible but it may also play a role in microtubule elongation. HMW MAPs have a similar action and it has been shown that they not only cause an increase in the initial rate of assembly, but also in the total amount of assembled microtubules

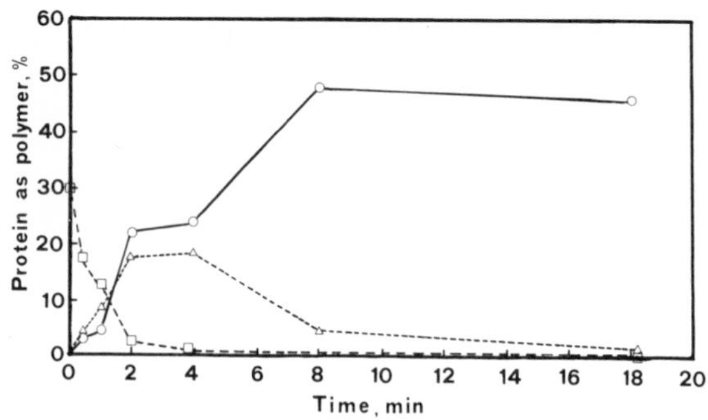

Fig. 3.8 Percentage protein found as rings (□), ribbons (△), and microtubules (O) during progress of polymerization. (Redrawn from [112].)

(elongation). There is evidence, however, that the latter activity may be more prominent [141]. Another possible function of HMW MAPs may be to form connections between microtubules and other constituents of the cytoskeletal system (i. e., microfilaments and intermediate filaments).

A promising approach to the study of microtubule formation and its control is based upon the use in *in vitro* systems of structures found within the cells, called microtubule organizing centres, which are able to promote microtubule assembly within the cell. Although not yet studied in detail, it would appear that, when isolated, these structures may also function in a similar fashion *in vitro*. This has been proved in the case of flagellar basal bodies and in the kinetochores of isolated metaphase chromosomes.

3.3.2 In vivo *microtubule assembly*

A fruitful method for studying some aspects of microtubule assembly *in vivo* (particularly the influence of different environmental conditions, such as temperature, pressure or cell cycle events) depends on the direct microscopic observation of the spindle formation during cell division. Using highly sensitive polarized optical systems, several workers have been able to directly observe microtubular structures by their birefringence, a phenomenon attributed to the bending of polarized light by microtubules. Important results have been obtained by Inoue and his group working with oocytes of marine echinoderms (for detailed information see [73]), as well as by other groups working with other systems.

It has been shown that chemicals (D_2O or certain glycols), or elevation in temperature, increase spindle birefringence (polymerization). On the other hand, low temperatures, high pressure and microtubule inhibitors, such as colchicine or vinblastine, decrease birefringence. Furthermore, the rate of microtubule assembly has been measured by following spindle growth (usually 1.5μm per minute).

It is not worthwhile discussing all the experimental data obtained, but one should refer to some important conclusions which have emerged from the results of the above examples and other *in vivo* studies. Two kinds of control regulate microtubule assembly within a cell: one is a temporal control that regulates the polymerization or depolymerization of tubulin at precise moments of the cell's life. The other is a spatial control necessary for the correct construction of microtubule-containing structures. There is evidence that in animal cells temporal control is not due to the amount of tubulin present in a cell, i. e., a cell does not start to assemble microtubules simply because the pool of its internal tubulin increases. The tubulin content within a cell is relatively constant throughout the cell cycle [124] and certain events connected with the assembly of large amounts of microtubules, such as neuronal growth in neuroblastoma cells, do not seem to need extra synthesis of tubulin [97]. Therefore, there must be other factors able to control the assembly of tubulin or the disassembly of microtubules when needed. Attempts have been made to identify these factors as being specific proteins, such as MAPs or a modification of the internal pool of Ca^{2+} ions, or more sophisticated control mechanisms mediated by cAMP, but up to now no clear evidence of how these factors may be involved has been obtained. In contrast, in some lower eukaryotes it seems possible that microtubule formation is dependent upon *de novo* tubulin synthesis. This may be the case in flagellar and ciliary regeneration in *Chlamydomonas* and *Tetrahymena* respectively, and it certainly occurs during flagellar growth in *Naegleria*.

A spatial control of microtubule assembly requires the presence within a cell of centres able to organize the assembly of tubulin and of systems able to control microtubule elongation, orientation and connections with other cell components. In other words, these two systems are needed to specify the pattern of the microtubular structure that is under construction. Nucleating sites for microtubule assembly, better known as microtubule organizing centres (MTOCs) [97], are widely distributed in all cells. Their morphological appearance is well documented, but unfortunately their molecular constitution is not well known. The mitotic centres (the spindle pole of mitotic spindle), the kinetochores of chromosomes, and in general all those structures from which bundles of microtubules originate must be regarded as MTOCs.

Apart from the control of tubulin assembly into microtubules there must also be some kind of mechanism which dictates the type of microtubule structure to be found, whether it is an axoneme, an axostyle or a mitotic spindle, etc. To explain how each of these patterns is formed, two possible mechanisms have been proposed [156] (Fig. 3.9).

One mechanism explains the pattern as being a result of the formation of links connecting already existing adjacent tubules. The type and the number of links formed are responsible for the final microtubule arrangement within the structure. This seems to be what happens in the axostyle assembly of certain heliozoans.

In the second mechanism, the pattern is determined by a pre-existing spatial arrangement of nucleating sites for each element of the structure, and therefore microtubules grow in prearranged positions. This probably occurs in ciliary and flagellar construction where the circular

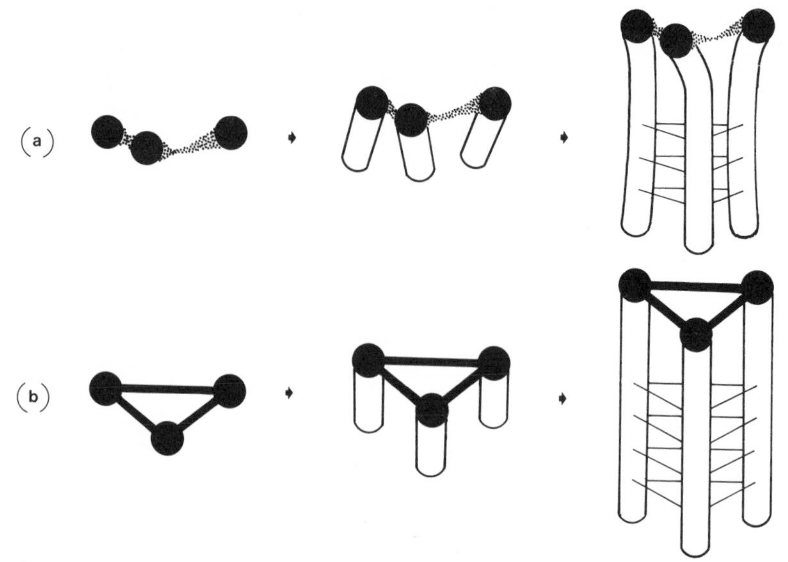

Fig. 3.9 Models of pattern formation in microtubule structures. In (a) pattern is specified after microtubules assembly; in (b) microtubules grow with a precise pattern defined by the basal structures. (From [165].)

arrangement of microtubules in the axonemes is established at the beginning of basal body morphogenesis, when nine tubule triplets, in a regular circular arrangement, start to assemble [165].

3.4 How microtubules can generate movement

Microtubules are able to generate movement by at least two different mechanisms: active sliding and changing their length.

The sliding filament model of ciliary and flagellar movement proposed by Peter Satir [128, 130] has received so much support from independent observations made in different laboratories that it can now be regarded as proved. The basic components of this model are microtubules and dynein arms, and the energy for displacement is derived from ATP hydrolysis.

Evidence that dynein (the ATPase protein which constitutes the lateral arms of flagellar and ciliary doublets) is directly involved in generating motility was first obtained from experiments with flagellar axonemes of sea urchin sperm. After selective trypsin digestion of spokes and nexin links, the axonemes with intact lateral arms were still able to move [152]. Further proof comes from the study of human patients suffering from a disease known as Kartagener's syndrome, in which sperms are non-motile because they do not possess dynein arms attached to the axonemal doublets [9, 111].

Filament sliding of axonemal doublets can be followed directly by dark field microscopy with axonemes partially digested by trypsin. Upon addition of ATP, filaments can be seen to become longer and thinner. This experiment shows that when adjacent doublets, unlinked

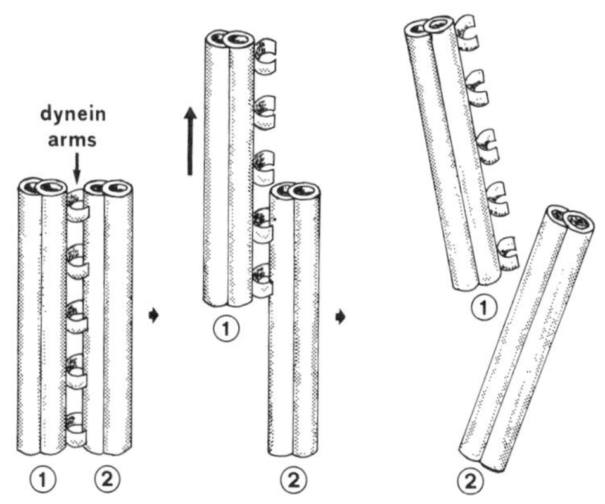

Fig. 3.10 Sliding mechanism in axonemal microtubules. Doublet 1 slides on doublet 2 by means of the movement of dynein arms.

by nexin, or radial spokes, which have been digested by trypsin, are free to slide against each other, they slide until separate. Therefore, what is the role of dynein arms in this model? Dynein arms are able to form fixed bridges between the A subfibre, from which they originate, and the B subfibre of an adjacent doublet. Upon hydrolysis of ATP, the bridge is broken and at the same time a new orientation of the dynein molecule occurs. This causes sliding. The repetition of this process causes a conspicuous sliding movement and can eventually result in the complete separation of adjacent filaments (Fig. 3.10). A good analogy is to imagine a caterpillar (doublet with dynein) crawling along a twig (adjacent doublet) by moving its legs (dynein arms) until it reaches the end of the twig and falls off. Sliding doublets connected by nexin links and radial spokes bend instead of sliding until separation, but this is another story and will be discussed in the section dealing with the mechanism of ciliary and flagellar movement.

Microtubules can also generate movement through the control of their assembly and disassembly. This type of movement can be exemplified by the extension and retraction of axopodia of heliozoans, and also probably by chromosome movement. Heliozoans have long linear appendages radiating from their bodies, giving them a 'sunny' appearance (the world *heliozoan* is derived from greek *helios*, meaning sun). The appendages have an internal skeleton of microtubules in a typical spiral arrangement. When an actinophryd heliozoan walks on a flat surface (Fig. 3.11), the attached axopodia protruding ahead of its body shorten by about 20 μm min^{-1}, while the attached axopodia behind lengthen by about 7 μm min^{-1}. These changes in length are respectively dependent on the depolymerization and polymerization of the microtubule axis, followed by a retraction of the axopodium, probably due to the surface tension of the cell membrane. In actinophryd heliozoans, the retractions are particularly slow since a large number of microtubules (hundreds) must be disassembled from the tip of the axopodium [169]. In some of the smaller heliozoans (centrohelids)

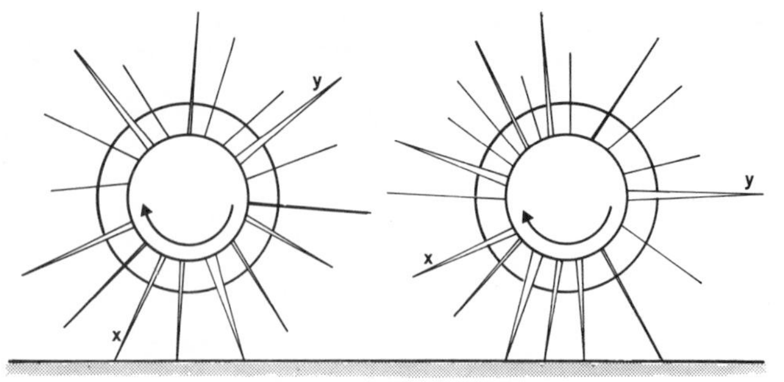

Fig. 3.11 Schematic diagram of the movement of an heliozoan by means of its axonemes.

axopodia contain only six microtubules and retractions occur at a very fast rate. This is also probably due to the fact that, upon application of a correct stimulus to the axopodia, depolymerization takes place along the whole length of the microtubules and the immediate collapse of the structure occurs (Fig. 3.12). Re-extension, however, is slower and takes place at the same rate as in actinophryd axopodia.

Although the molecular mechanism of chromosomal separation is at the moment most controversial, much evidence indicates that assembly and disassembly of spindle microtubules may be involved. More or less satisfactory models of chromosome separation exist, but a general agreement has not yet been reached and the discovery of acto-myosin filaments in mitotic spindles has added further confusion. For this reason it is better to postpone the argument until further research has provided more conclusive data.

3.5 The problem of intermediate filaments

The presence of filaments of about 10 nm in diameter (i. e., narrower

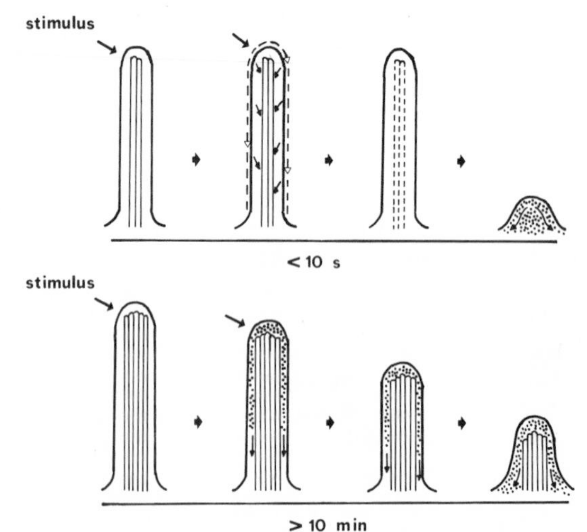

Fig. 3.12 A hypothetical comparison of the retraction of axopodia following appropriate stimulation of a *centrohelid* (top) and an *actinophryid* (bottom) heliozoan.

than microtubules and larger than microfilaments) as single elements or in small bundles in the cytoplasm of cells has been recognized since the early use of electron microscopy in biology. The relative amount of these structures varies according to the cell type examined: in epidermal cells they are widely distributed and are known as *tonofilaments*. In other cell types, such as cells of the central nervous system, they are called *neurofilaments* and are present in conspicuous amount in axons. In other mammalian cells their amount is not particularly relevant but they are still found [52].

Although the chemical composition of intermediate filaments is not yet clear, they appear to be constituted of one or more major proteins. In baby hamster kidney cells (BHK-21) intermediate filaments have been resolved, using polyacrylamide gel electrophoresis, into two bands with molecular weights remarkably similar to that of tubulin subunits [145]. There is evidence, however, that they constitute a homogeneous class of proteins distinct from other cytoskeletal proteins. Antibodies obtained against intermediate filaments of one cell type cross-react with filaments of other cellular types, but do not with microtubules and microfilaments.

Intermediate filaments have been briefly mentioned in this book because there is evidence that: (a) these structures may be physically associated to microtubules, and (b) they may play a role in motility in mammalian cells. Evidence supporting the former comes both from electron microscopical observation of fine bridges of fibrillar material which connects the microtubule wall to intermediate filaments, and from the fact that antimitotic drugs (such as colchicine), induce redistribution of intermediate filaments *in vivo*. Support for the second point comes from the electron microscopical evidence of a different location of intermediate filaments in the cytoplasm of leukemia cells during locomotion [45]. More work, however, is necessary before the characters and the functions of this most unexplored part of the cytoskeletal system becomes clear.

3.6 Microfilaments

Microfilaments of eukaryotic cells are long thread-like structures 5–7 nm wide, present in the cytoplasm of cells. From a chemical point of view, they appear to be mainly constituted of actin (although other proteins can also be found) and are therefore frequently known as actin filaments. Microfilaments may be found associated in groups or bundles, and in some highly specialized cells (such as muscle cells) they can form well-defined and stable structures. More frequently, however, they make up unstable bundles or fine networks that change their appearance and topographical location within the cell according to the cell cycle, cell movement, etc. This means that, like cytoplasmic microtubules, microfilaments of non-muscle cells are transient structures formed by molecules able to assemble and disassemble according to the needs of the cell in which they are found.

The morphological characters of microfilaments and their behaviour

State myosin is microfilament.

during cell movement and cell cycle have been extensively studied on cultivated animal cells using fluorescent antibodies against actin and other proteins, such as myosin, tropomyosin and α-actinin, which are frequently found in microfilaments [84, 86]. When cells are observed in resting phase, attached to the surface of the flask in which they live, long bundles of microfilaments seem to cross the cell, apparently just below the cell membrane, generally avoiding the ruffled edges of the cell. The bundles of microfilaments, called stress fibres, are characteristic of normal cells and disappear in cells transformed by oncogenic viruses. When cells are in movement, a rather different pattern is seen: in addition to stress fibres, fine bundles of actin filaments are dispersed in the cytoplasm and are particularly evident in the advancing projection of the cells. During cell division fluorescence is associated with the mitotic spindle and the images obtained are remarkably similar to those obtained with anti-tubulin antibodies under the same conditions. At cell separation microfilaments are condensed in the bridge region between the dividing cells where they form a contractile ring, which is responsible for cell separation (Fig. 3.13). From this description it is evident how the microfilament arrangements in cultivated cells may vary according to the cell cycle.

There are other cells, however, in which microfilaments are stable organized structures and do not undergo frequent changes in shape and distribution. Apart from muscle cells (which are specifically treated in another book of this series and therefore will not be considered in our discussion), a good example of a stable organized microfilament structure is found in the long finger-like protrusions known as

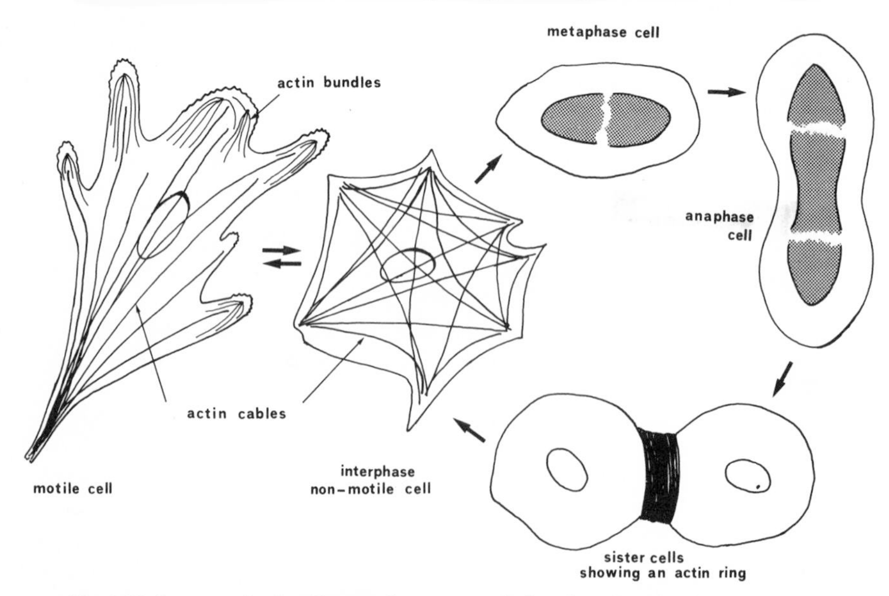

Fig. 3.13 Patterns of microfilament fluorescence during the cell cycle, obtained with fluorescent-labelled antibodies against actin.

microvilli, which extend from intestinal epithelial cells, and many other animal cells. Within each microvillus a bundle of microfilaments runs parallel to the axis of the microvillus, being attached at the tip and along the axis to the cell membrane [100]. Other well-studied examples of organized bundles of actin are in the microfilament bundles of echinoderm sperm and sea urchin oocytes [157]. By applying techniques of optical diffraction and three-dimensional reconstruction to *Mytilus* sperm and sea urchin oocyte microfilaments, it has been shown that within each bundle microfilaments are regularly packed into hexagonal arrays and are connected by regular bridges in a paracrystalline fashion [37]. The organization of this structure is remarkably similar to the regular arrangement of microtubules in the axostyle of flagellated protozoa.

Seen under an electron microscope, actin filaments of non-muscle cells are rarely found in association with the thicker and shorter filaments of myosins (8–9 nm wide and 300 nm long) the proteins which interact with actin for the generation of movement. The lack of electron microscopically-evident myosin filaments in non-muscle cells is probably due to their size, which makes them difficult to see in a thin section. In fact, when microfilaments are stained with antibodies against myosin, a diffuse fluorescence is seen, demonstrating that myosin is indeed present. Fluorescent antibodies produced against other proteins of the contractile system also decorate cytoplasmic microfilaments with typical patterns: e. g., anti-α-actinin and tropomyosin antibodies show a fluorescence in alternating patches along microfilaments, demonstrating the presence of these two proteins in association with them.

Essentially, two functions have been attributed to microfilaments in non-muscle cells. Their classical role is as generator of motility; they are therefore regarded as being responsible for amoeboid movement. Through contraction they can provide the forces for other cellular activities, such as changes in cell shape, cytoplasmic streaming, movement of organelles in the cytoplasm, phagocytosis, cell secretion, cell division and the regulation of topographical distribution of proteins in cell membranes. There is increasing evidence, however, that they can also play a cytoskeletal role by providing rigid bundles of filaments which may help the cell to maintain a definite shape [32, 80].

3.7 Molecular components of microfilaments

In order to understand better the dynamics of microfilaments and their function in cell movement, the most relevant biochemical characters of the microfilament components in non-muscle cells will be inspected.

3.7.1 Properties of non-muscle actins

Actin is a protein widely distributed in all the eukaryotic cells examined so far. It can be found in soluble monomeric form or as assembled filaments (Fig. 3.14). In some actively motile cells, such as amoebae, macrophages or platelets, it is the predominant protein of the cell extract and may constitute 20–30 per cent of the total protein content. In other less motile cells it represents a rather small fraction of the total protein

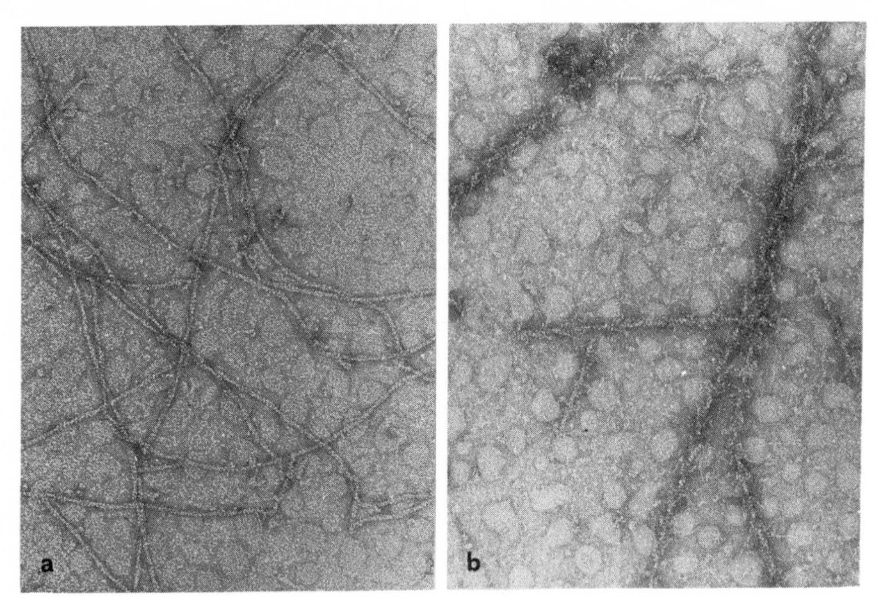

Fig. 3.14 Electron micrograph of microfilaments formed *in vitro* by calf brain actin. In (a) normal microfilaments; in (b) after reaction with heavy meromyosin that binds specifically to actin microfilaments. Negative staining, × 120 000. (Courtesy of Dr. G. Monaco.)

content, being as little as 1–2 per cent. However, this is still remarkable if one considers that in a eukaryotic cell many thousands of different proteins may be present at the same time.

Due to its relatively high concentration, it has been possible to extract actin from cell systems. About thirty different actins from different cell types (varying from slime mould amoebae to man) have been studied to varying extents.

Actin purification can be carried out with the usual techniques of protein separation, involving ion exchange chromatography and gel filtration [144]. It can be obtained more simply and with a reasonable degree of purity by affinity chromatography of high-speed supernatants of whole cell extract on agarose beads to which the enzyme DNAase 1 is covalently attached. Actin has a great affinity for DNAase 1, is retained by the agarose beads, and can be released upon addition of the appropriate buffers [85]. The affinity to DNAase 1 is a well-known characteristic of the actin molecule: the protein binds tenaciously to the enzyme giving rise to inactive DNAase 1–actin complexes. However, the physiological significance of this reaction is unknown [68].

All the actins so far examined show remarkable similarities with respect to molecular weight (always around 42 000 daltons), electrophoretic mobility, amino acid composition (all posssess the unusual amino acid N-methylhistidine, whose function within the molecule is unknown), the ability to bind adenine nucleotide at equimolar amounts, and the ability to form polymers [80]. Amino acid sequences of actins from different sources show, however, the existence of small but important differences, sometimes of only one amino acid. Differences have also been found in the chemical composition of actins from different organisms, and within a single organism among different tissues. Furthermore, it has been shown that within a single cell different types of actin may coexist. Their dissimilarities are very subtle indeed and can be found only by coupling two powerful separating techniques, such as isoelectric focusing (which allows separation of proteins according to their isoelectric point) and SDS-polyacrylamide electrophoresis (in which separation is due to the molecular size). In this way three types of actin have been identified and have been called α-, β- and γ-actin respectively. Two additional unstable forms may also be present. The α-form of actin is found in fully mature muscle tissue; β- and γ- actins coexist within most of the non-muscle cells examined; in cultivated myoblasts all three types of actin can be found. Since the mRNAs translating these three types of actin have also been isolated, it can be concluded that within a single cell at least three different actin genes exist and their expression is strictly controlled in each cell type [70, 151].

3.7.2 Proteins interacting with actin

Actin is known to interact with many other proteins. According to the proteins which are involved as a result of the interaction, many events can occur. These can be the generation of motility, the control of polymerization of actin, the establishment of connections, and possibly the regulation of some steps in metabolic pathways.

The protein which is most frequently found in association with contractile filaments is myosin. The characteristics of myosin from muscle cells have been studied extensively, but less is known about it in non-muscle cells. Myosin can be identified by its adenosine triphosphatase enzymatic activity, usually stimulated by actin, and by its ability to interact with actin [32]. The stimulatory effect of actin is related to the presence of particular ions, and therefore myosins can be categorized according to their dependence on Ca^{2+}, Mg^{2+}, or K^{2+} ions. From a structural point of view, most of the non-muscle myosins so far examined consist of two heavy polypeptide chains of about 200 000 daltons each and two pairs of light chains of about 17 000 and 20 000 respectively. The resulting molecule is in the range of 500 000 daltons and consists of two distinct regions. The so-called 'rod region' is a double-stranded α-helical structure formed by the combination of part of the heavy chains, and is able to react with other myosin molecules forming bipolar filaments (Fig. 3.15). At one end of the rod region are located two heads, with the actin binding-site and the ATPase site present in each. Light chains are found to be associated with the head-region.

A myosin which has a structure different from that depicted above is myosin I, found in the soil amoeba *Acanthamoeba castellani.* This protein is remarkable because it contains only one heavy chain of

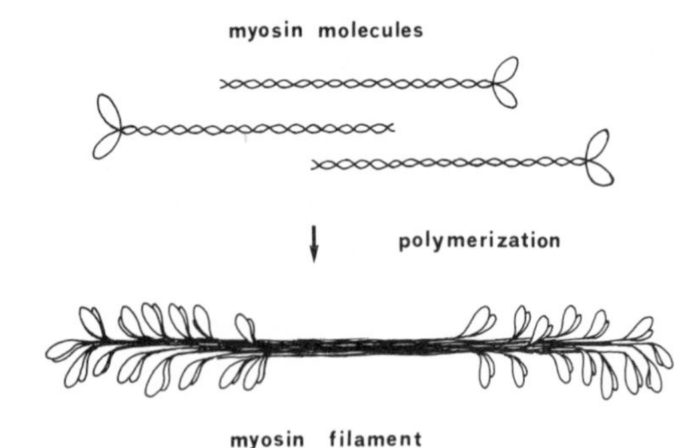

Fig. 3.15 The polymerization reaction of myosin molecules to form bipolar filament.

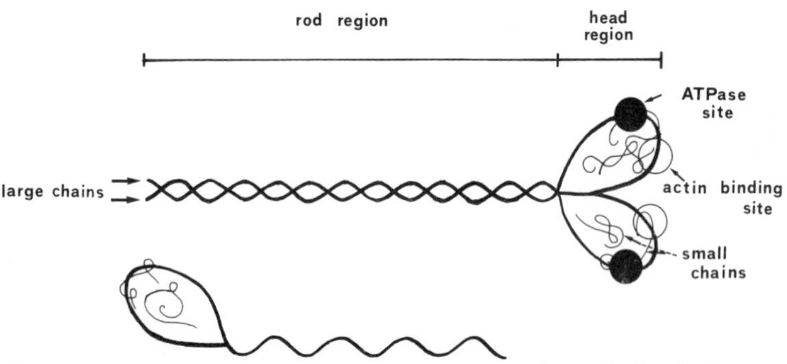

Fig. 3.16 Schematic representation of a normal myosin molecule (top) and of mono-headed myosin from *Acanthamoeba* (bottom).

140 000 daltons and two light chains of 14 000 and 16 000 daltons respectively (Fig. 3.16). It cannot form bipolar filaments and is activated by actin only in the presence of a cofactor [118, 119]. From the same amoeba, another myosin (myosin II) has recently been isolated. This more typical myosin is able to form bipolar filaments and has two heads showing ATPase activity [120].

The Mg^{2+}-dependent ATPase activation by actin in some non-muscle myosins requires a previous phosphorylation of a light chain by a specific protein-kinase which is able to transfer a phosphate from a donor to the 20 000 light chain. This is also true for *Acanthamoeba* myosin I and recently its co-factor has been identified as a kinase that specifically catalyses the phosphorylation of the heavy chain [95]. From a genetic point of view, the myosin system is not so well conserved as the actin one, as so far about fifteen types of molecules have been studied, revealing many different characteristics.

Apart from myosin, molecules interacting with actin or actomyosin filaments can be divided according to their activities into broad classes (Table 3.2). For some of them, such as tropomyosin and α-actinin, their presence in association with microfilaments has been directly visualized using immunofluorescence [84].

Tropomyosin isolated from mammalian platelets, brain, pancreas

Table 3.2 Some of the proteins known to interact with actin in non-muscle cells

Source	*Interacting proteins*		*Interacting proteins*	*Source*
			Filamin	Macrophages
			α-actinin	Different systems
Spleen	Profilin	G-actin ⇄ F-actin	Gelactins I-II-III-IV	*A. castellani*
Pancreas	DNAse		Troponin	Different systems
			Tropomyosin	Different systems
			Actin-binding protein	Macrophages

and fibroblasts has a molecular weight and a peptide map similar to that of muscle tropomyosin and is involved in regulating the Ca^{2+} concentration necessary for actomyosin contraction.

α-actinin, or similar proteins, have been found particularly in intestinal microvilli, in cultivated animal cells and in the axonemal process of horseshoe crab sperm. It seems somehow to link microfilaments to cell membranes, and perhaps to other structures, and is present mainly in those areas where these contacts are made.

Other proteins are able to inhibit actin polymerization in filaments *in vitro* and therefore may act to control the formation of filaments *in vivo*. Among proteins which prevent actin polymerization, one, called profilin, has been purified from spleens and characterized as a 16 000 daltons polypeptide able to form a complex with soluble actin. Similar proteins are probably widely distributed among non-muscle cells, particularly in those in which a high concentration of soluble actin is present.

Certain proteins, such as filamin and the actin-binding protein isolated from macrophages, are able to bind to the actin microfilament *in vitro* and thereby prevent actin from stimulating the ATPase activity of myosin, resulting in inhibition of acto-myosin–dependent contraction. Their *in vivo* function is unclear, but these proteins may be used by the cell to restrict the filament to a simple cytoskeletal role [80]. Other proteins that may have similar functions are the gelactins, a group of four different proteins isolated from *Acanthamoeba*, which are able to gelify solutions containing microfilaments by forming bridges between them. These proteins also inhibit myosin Mg^{2+}-dependent ATPase in a way resembling the action of filamin and the actin binding protein.

Finally, among all the proteins which interact with actin, there are also aldolase and several other enzymes of the glycolitic pathway. This interaction is probably not relevant for microfilament function but it may indicate an *in vivo* regulatory function of actin on certain metabolic pathways.

The system described above is already complicated yet the studies done so far have only scratched the surface. The many ideas and theories about actin and associated proteins in non-muscle cells require further study before any concrete conclusions can be made.

Fig. 3.17 The polymerization reaction of actin subunits into microfilaments.

3.8 Microfilament assembly and its control

As we have already mentioned, actin exists in two functional states: a soluble monomeric state and a polymeric state resulting from the assembly of single subunits into double stranded filaments. Monomeric actin is known as G actin, and the polymeric form as F actin. Double-stranded helices of F-actin are about 7 nm wide and show a half-pitch of 34 nm (Fig. 3.17).

The *in vitro* reaction of polymerization occurs at an optimal salt concentration and in the presence of bound ATP. The actin concentration is critical for the reaction, and the polymer formed will always be in equilibrium with the monomer at its normal critical concentration, according to the equation:

$$(\text{actin})_n \rightleftarrows (\text{actin})_{n-1} + \text{actin}$$

The critical concentration at which an actin polymerizes in a specific solution is specific for each type of actin and can be used to define it.

Since the concentration of unpolymerized actin in many cell types is largely above its critical concentration, there must be factors able to interact with actin preventing its polymerization. These factors must therefore by regarded as regulators of the filament formation. Protein inhibitors of actin polymerization have been found in many cell types and one of them (profilin) has already been mentioned. As an example of a system in which inhibitors of actin polymerization are intimately involved, we shall examine the sperm cells of echinoderms such as the sea cucumber (*Thyone*). These cells are also a remarkable example of polarized assembly of microfilaments and will therefore be discussed in some detail. The head of a *Thyone* sperm has an apical end with a cup-shaped depression in which is contained an acrosomal vacuole. Between the acrosomal vacuole and the chromatin material of the nucleus lies an amorphous material which is actin in a monomeric state. When a sperm approaches an egg cell of the same species, the acrosomal reaction occurs within a few seconds of contact of the acrosomal tip with the extracellular material around the egg. This involves the rapid formation of a thick bundle of actin microfilaments which originates from the bottom of the cup and protrudes from it by about 90 μm (Fig. 3.18). The function of this new formation is to penetrate through the jelly material that surrounds the egg cell, allowing the fusion of the two cell membranes. From this description it is evident that, before activation, the head contains a high amount of unpolymerized actin, enough to make up a 90 μm long bundle of microfilaments, compressed into the small cup-shaped area. It is therefore highly possible that proteins which

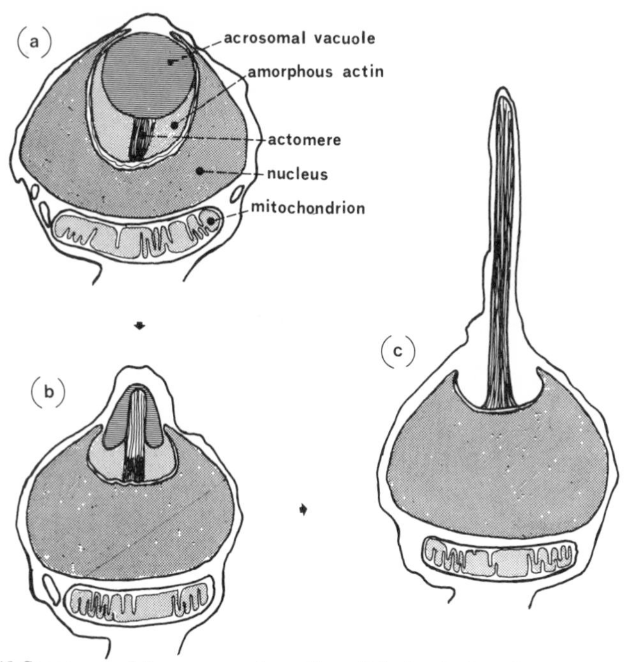

Fig. 3.18 Sequences of the acrosomal reaction of the head of a *Thyome* sperm (a), showing the beginning of actin polymerization in (b) and the completed acrosome in (c).

prevent actin polymerization could be found there. In the search for these proteins actin material of isolated *Thyone* sperm heads has been subjected to SDS acrylamide gel electrophoresis. Three major peptides were found, one corresponding to actin. The other two, with a molecular weight of 230 000 and 250 000 daltons, are probably the proteins responsible for the maintenance of actin in the unpolymerized state. By treatment with the proteolytic enzyme trypsin, monomeric actin may be obtained free from the high molecular weight components [158]. The acrosomal reaction of *Thyone* sperm can be induced by one of many activators, and requires an increase of the internal pH which frees the actin from binding proteins, leaving it able to polymerize. Polymerization is initiated by a specific structure that can be seen at the bottom of the acrosomal cup, and has been called the *actomere*. Morphologically it is composed of a bunch of short microfilaments (about 25) embedded in a dense amorphous material. During microfilament formation the actomere behaves as a nucleating body from which microfilaments grow [160].

According to this and other evidence, it has been proposed that microfilament polymerization and growth is controlled by organizing centres that act similarly to MTOCs for microtubules.

Specific sites for growth of actin filaments have also been demonstrated in *Mitylus* sperm, *Limulus* sperm and in the formation of intestinal microvilli [150] and it may be possible that they exist in many other situations. In fact, microfilaments are frequently seen attached to specific sites of non-muscle cell membrane, but at present it is unclear whether these sites control the assembly of microfilaments or whether the attachment to them occurs after assembly has taken place.

3.9 How microfilaments generate movement

Microfilaments can generate movement in two different ways: by a sliding mechanism in which actin and myosin filaments slide against each other, or simply by assembly and disassembly of microfilament bundles. Here again a remarkable correspondence between the microfilament and microtubule systems can be found: i. e., both system generate movement in the same way. This is interesting from an evolutionary point of view because it indicates that systems which presumably have evolved separately have resolved their problem of generating movement in an identical way, although with different components.

The sliding mechanism model for microfilament motility is derived directly from the study of muscle contraction and, perhaps with minor modifications, can be used to explain acto-myosin-dependent movements in non-muscle cells.

The system is basically composed of actin microfilaments which have one end attached to various structures of non-muscle cells, such as cell membranes, microtubules or cell organelles, and the other end is left free. Between two opposing free ends of actin microfilaments are bipolar filaments of myosin. When the two opposite actin filaments slide on myosin, their free ends approach each other while the other ends pull along the structures to which they are attached. In muscles these events occur in a structure specifically constructed to generate motility (the sarcomere), composed of sets of microfilaments attached to Z lines on both sides and interlocked by myosin filaments. When movement occurs, it reduces the length of the sarcomere, which corresponds to muscle contraction. When similar events occur in non-muscle cells, the result is the movement of the structures to which microfilaments are attached (Fig. 3.19).

At this point it seems worthwhile to explain how sliding can be generated by the interaction of actin filaments with myosin. The true effector of movement is myosin (more specifically its head, where the ATPase activity is located). Myosin heads show a great affinity for ATP and bind one molecule of ATP to each head, when ATP is available. As soon as the ATP binds, the myosin head is transformed into an activated form with a high affinity for actin and it therefore attaches itself to an actin subunit in the nearest microfilament. However, the attachment to actin causes the immediate hydrolysis of the ATP and the energy released is used to rotate the head by a small angle, which slightly moves the actin filament to which it is attached. In the presence of fresh ATP the cycle is repeated many times and therefore a conspicuous sliding can result. This is a very simple description of the events which are responsible for acto-myosin-dependent motility in muscle cells (a

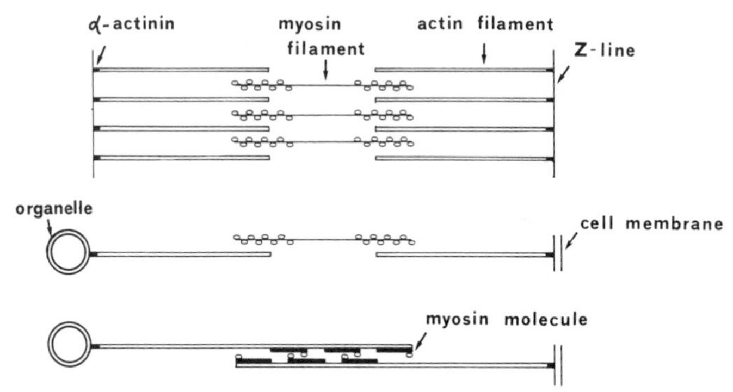

Fig. 3.19 Models of acto-myosin interaction in muscle (top) and non-muscle cells (middle and bottom).

more complete description can be found in *Muscle Contraction* by R. M. Simmons, another book in this series). This basic mechanism of acto-myosin interaction can probably be transposed to non-muscle cell systems as well.

3.10 Sliding control in microfilaments

In muscle cells the sliding reaction is made sensitive to low levels of Ca^{2+} ions by the presence of two proteins, tropomyosin and troponin, which bind to the actin filament. The binding of the ATP-myosin complex to actin can occur only in the presence of Ca^{2+}, and therefore Ca^{2+} is a regulatory molecule for muscle concentration. Its concentration in the tissues is regulated by its release from the sarcoplasmic reticulum as a consequence of membrane depolymerization [41].

If we want to transpose the events of muscle contraction onto non-muscle cells we can imagine that an interaction between the actin filament and bipolar myosin filaments occurs in a similar way. Unfortunately, bipolar myosin filaments cannot be demonstrated in non-muscle cells in appreciable amounts, except in particular cases such as in platelets. However, they may be present but simply not seen under the electron microscope because of their size and their scattered arrangement. In fact, animal cells stained with anti-myosin fluorescent antibodies demonstrate that myosin is indeed present in microfilaments [48]. It is also possible that in non-muscle cells myosins do not arrange themselves into classical multiheaded filaments but in a more simple double-headed form still able to interlock actin microfilaments. Another model put forward by Maruta and Korn [95] proposed that single molecules of myosin could be attached through the rod segment of their heavy chain to the actin filament in a way that leaves their heads free to interact with other nearby filaments. In this case motility would be generated by direct sliding of two filaments against each other, a mechanism occurring among flagellar and ciliary microtubules. According to this model, one-headed myosin that apparently cannot form bipolar filaments (such as the *Acanthamoeba* myosin I), could also have a function in generating motility (Fig. 3.19).

Since motility depends on the interaction of actin with myosin, factors that control this interaction may also be regarded as controllers of cell motility. In muscle cells the principal mechanism of control involves Ca^{2+} and the proteins tropomyosin-troponin that are associated with the actin filament. In non-muscle cells the regulatory systems are not well known. It is clear, however, that many regulatory systems may exist in different types of cells, some of which are based on Ca^{2+} and others on different mechanisms.

Although the enzymic interaction between purified actin and myosin in non-muscle cells is not generally modified by Ca^{2+} levels, there are examples of Ca^{2+} sensitive ATPase activity in actomyosin from brain, leukocytes, platelets and the acellular slime mould *Physarum polycephalum* [32]. Ca^{2+} sensitivity can also be assessed by recording the contraction of acto-myosin filaments in cell extracts (this has been done for amoebae and other cell types). Evidence for *in vivo* Ca^{2+} sensitivity of microfilament-mediated motility has been shown in cytoplasmic streaming in large amoebae (*Amoeba proteus* and *Chaos carolinensis* and in *Physarum*) and in the ATP-dependent contraction of isolated brush borders of intestinal epithelium [98, 156].

In muscle cells the calcium control of contraction operated in the presence of the Ca^{2+} binding protein of the tropomyosin-troponin system, and therefore some investigators have looked for similar proteins in non-muscle cells. Tropomyosin-like proteins have indeed been found in platelets, brains, pancreas and mouse cultivated fibroblasts, and a Ca^{2+} binding protein similar to muscle troponin has recently been isolated from chick embryo brains. Other proteins which give Ca^{2+} sensitivity to the acto-myosin complex have been isolated from *Physarum* and *Dictyostelium*, but need further study.

Apart from the Ca^{2+} control of microfilament contractility, other systems for the regulation of acto-myosin interaction do exist.

The control of myosin phosphorylation can be regarded as one of these. It has been shown, for instance, that the actin-stimulated ATPase activity of myosin in platelets increases about five times when the 17 000 dalton light chain of myosin is phosphorylated by a specific protein kinase in the presence of ATP [4]. This suggests a direct control of phosphorylation over acto-myosin interaction. However, it is not yet possible to draw final conclusions about this system.

It seems quite evident from these examples that, at the moment, a unified theory of the regulation of acto-myosin interaction in non-muscle cells does not exist. This may depend partly on the relatively scant amount of data available on the subject and partly on the real complexity of the problem. It is likely that many systems operate in controlling acto-myosin interaction in non-muscle cells. The control of the concentration of Ca^{2+} ions is important for certain types of cells, but little is known about the mechanisms operating in other systems.

The other way of producing motility, probably not for the cell as a whole but instead for some parts of it (such as the membrane), is by polymerization and depolymerization of bundles of actin filaments. In this case, movement is not generated by sliding of microfilaments but is directly due to the growth of filament bundles which push apart whatever is in contact with their growing fronts (the reverse may be imagined when disassembly occurs).

An example of how this type of movement operates occurs in the above-mentioned acrosomal reaction. During this reaction, within ten seconds a straight filament bundle assembles, thereby causing the protrusion of the sperm membrane toward the egg.

In the cytoplasm of some non-muscle cells are frequently found other types of supramolecular assembly of actin monomers, not in the form of microfilament bundles but as a fine network of fibres. These phenomena can be reproduced also in *in vitro* systems and are known as the reaction of gelation. They are probably more important for controlling cytoplasm viscosity and cell shape, and although these functions may be indirectly connected with cell motility, they cannot be considered as true motile events [80].

3.11 Microtubules, microfilaments and cell membranes

There are many reasons that lead one to think of cytoplasmic microtubules and microfilaments as highly interconnected structures of the same system. Both have similar roles in generating motility and, as cytoskeletal structures, they inhabit the same cell, behave in a similar way according to the cell cycle, and form structures morphologically similar. It is unpleasant to admit, however, that despite these considerations, direct proof of an interconnection between microtubules and microfilaments does not exist. There is indeed much indirect evidence obtained using different methods which makes the presence of a connection between the two systems highly probable, and it is almost universally accepted that such a connection does exist.

The evidence comes from morphological studies of microtubule and microfilament distribution within the cell, particularly during mitosis. Fujiwara and Pollard [49], using anti-tubulin and anti-myosin antibodies conjugated with fluorescein and rhodamine respectively, were able to compare the patterns of microtubules and microfilaments within the same cell. In fully spread cells they found no morphological association of the two classes of fibres. In the mitotic spindle, however, myosin fluorescence was seen associated with microtubule fluorescence in a way that could indicate a close relationship between the different fibres.

More indirect evidence is derived from the use of inhibitors of microtubule structures, such as colchicine or its derivative colcemid, and inhibitors of microfilaments such as cytocalasin B [166]. As an example of this approach we can consider the influence of microtubule and microfilament inhibitors on the 'capping' phenomenon of mammalian cells [40, 104]. The 'capping' reaction occurs when a cell is treated with divalent antibodies or with other ligands able to interact with proteins on its membrane. It consists of the condensation of the linked molecules in a defined region of the cell surface. The 'capping' reaction can be

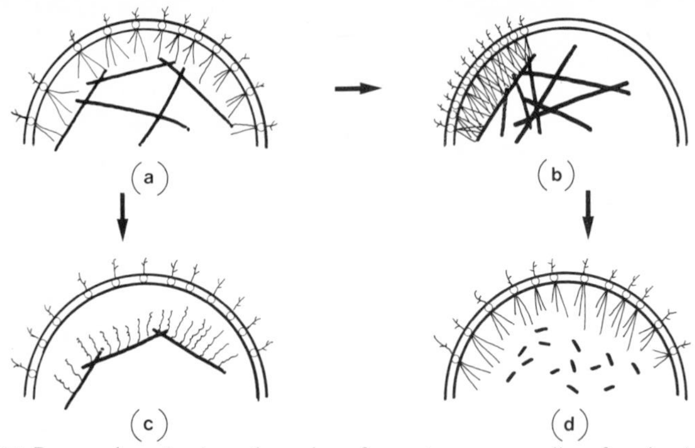

Fig. 3.20 Proposed mechanism of capping of receptors over a cell surface (a – b) and the reversal due to microfilament (c) and microtubule(d) disrupters.

inhibited by cytocalasin B, suggesting that microfilaments play an active role in the movement of molecules on the cell membrane [104]. This inhibition of capping is in many systems reversed by treating with colchicine or with other disrupters of the microtubular system [153]. When microtubules and microfilaments are disrupted by a combination of colchicine and cytocalasin B, capping, usually a stable event, can be reversed and a new redistribution of the capped molecules occurs. These results have suggested that both microtubules and microfilaments are involved in the control of movement of molecules in the cell membrane: microfilaments providing the contractile activities and microtubules serving as a skeletal anchorage system for microfilaments. They also suggest that a direct connection between microfilaments and cell membrane proteins is necessary for lateral movement of proteins in the lipid bilayer (Fig. 3.20).

There is, however, more direct evidence that microfilaments are connected to surface proteins and therefore to cell membranes. Several researchers, using different techniques, have found an accumulation of microfilaments in the region which lies immediately beneath the antibody induced in lymphocytes or in the concanavalin A-induced cap in ovarian granulosa cells [10]. In the same area there is an accumulation of myosin and tubulin. More recently, using a technique for precipitating actin from crude cellular lysate by interaction with muscle myosin, followed by the identification of the molecules which coprecipitate with actin, it has been shown that actin precipitates with molecules which are expressed on the external surface of cell membranes, such as H-2 antigens and immunoglobulins. The amount of immunoglobulins found associated with actin is greater in cells which have been capped than in those with an undisturbed distribution of surface antigens [46]. According to these results, the aggregation of surface proteins occurring in capping could be coupled to the formation of new links with actin. This provides further support for the idea that microfilaments control at least certain types of motility of membrane components.

The above observations also demonstrate a direct association between actin filaments and membranes. Evidence obtained with other methods and in other systems further supports this hypothesis. One of these derives from the frequent isolation of microfilaments during the process of purification of the cytoplasmic membranes. This was first shown for *Acanthamoeba castellani* [117], but was later recognized in almost all types of cells. Microfilaments of purified membranes can be observed under an electron microscope (and they appear to be attached to the internal surface), and actin can also be detected using electrophoresis in SDS.

Further evidence comes from the direct microscopic observations of microfilaments in areas of the membrane where they are found in high quantities. Sometimes, as in the case of microvilli, a clear image of microfilaments attached to the membrane at the tip of microvilli is seen. These spoke-like connections between the central bundle of filaments and the microvilli membrane have been observed in electron micrographs of cross-fractured microvilli.

How the microfilaments are attached to the membrane is still unclear. Is there a direct attachment of microfilaments to membranes or is their union mediated by other proteins? The second hypothesis seems to be the most likely α-actinin or similar proteins are the most probable candidates. An α-actinin has been shown to be present at the microvilli tip where filaments are attached to the membrane and is found in the plasma membrane or in the membranes of secretory vescicles close to the microfilaments [80]. It has therefore been proposed that this protein has the same role in muscle cells (where it attaches microfilaments to the Z line of the sarcomere) and in non-muscle cells. Microfilaments could also interact with membranes through myosin molecules, and there is evidence which indicates that myosin may be associated with membranes as a transmembrane protein.

Less convincing experiments have been performed in trying to prove a direct interaction between microtubules and membranes. Colchicine binding to membranes has been reported by many researchers but, due to the various factors which may produce a positive reaction, this cannot be seriously used to prove the existence of tubulin in cell membranes. More direct evidence indicating tubulin as a membrane component has been found in brain synaptosomes [167] and in membranes of molluscan cilia [147]. In any case, the association of tubulin with membranes appears not to be as widespread as that of actin.

Having rapidly reviewed the evidence for connections among microtubules, microfilaments and membranes, and remembering the topographical distribution of the components of the cytoskeletal system in the animal cell, a model showing the connections between the components of the cytoskeletal system and the cell membrane can be formed (Fig. 3.21). According to this model, in interphase cells microtubules form an internal frame-like structure, radiating toward the cell periphery. This structure does not directly promote movement but may serve as a support for the more peripheral microfilament structure. Microfilaments, connected with membrane proteins and organelles, are

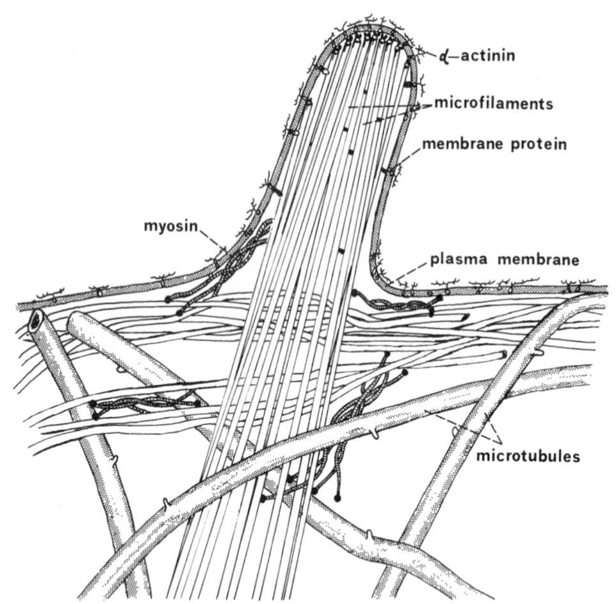

Fig. 3.21 Proposed relationships between microtubule, microfilament and cell membrane. (From Nicholson *et al.*, 1977, in *Dynamic Aspects of Cell-Surface Organization, vol. 3, North Holland.)*

responsible for movement of the intracellular structures and the cell as a whole. This model implies a cytoskeletal role for microtubules and a dynamic role for microfilaments.

Topics for further reading

Books of general interest on arguments treated in this chapter are:

Primitive motile systems in cell biology (1964) (ed. R. D. Allen and N. Kamiya), Academic Press, New York and London.

Poglazov, B. F. (1966), *Structure and function of contractile proteins,* Academic Press, New York and London.

Cell motility (1976) (ed. R. Goldman, T. Pollard and J. Rosenbaum), Cold Spring Harbor Laboratory, Cold Spring Harbor.

In order to understand the tremendous advance in the molecular aspects of cell motility during the last ten years it may be useful to compare the content of the first two books with that of the third.

Among the excess of reviews and specific articles on microtubules and microfilaments, the reader may find additional information for microtubules in:

Snyder, J. A. and McIntosh, R. (1976), Biochemistry and physiology of microtubules, *Ann. Rev. Biochem.*, **45**, 699–720.

Gaskin, F. and Shelanski, M. L. (1976), Microtubules and intermediate filaments, in *Essays in Biochemistry* (ed. P. N. Campbell and W. N. Aldridge), vol. 12, Academic Press, New York and London, pp. 115–46.

Stephens, R. E. and Edds, K. T. (1976), Microtubules: structure, chemistry, and function, *Physiol. Rev.*, **56**, 709–77.

Dustin, P. (1978), *Microtubules*, Springer-Verlag, Berlin, Heide
York.

And for microfilaments and related topics in:
Clarke, M. and Spudich, J. A. (1977), Nonmuscle contractile prote[illegible] of actin and myosin in cell motility and shape determination, *Ann. Rev. Biochem.*, **46**, 797–822.
Hitchcock, S. E. (1977), Regulation of motility in nonmuscle cells, *J. Cell Biol.*, **74**, 1–15.
Korn, E. D. (1978), Biochemistry of actomyosin-dependent cell motility, *Proc. Natl. Acad. Sci. USA*, **75**, 588–99.

4 The movement of eukaryotic cells

Microtubule- and microfilament-dependent motility in eukaryotic cells is a phenomenon that can be expressed in a number of ways, resulting in either the movement of the cell as a whole or only a part of it. In the first case an active locomotion is generated, in the second the movement is localized to a specific part of the cell, such as the membrane, the cytoplasmic organelles or the cytoplasm itself. Cell locomotion depends on two principal types of movement: the ciliary or flagellar movement and the amoeboid movement. Cilia and flagella of eukaryotic cells are cylindrical organelles, which when animated, propagate waves resulting in the movement of the cells, which are free to move. In fact, when cells are not free to move, ciliary and flagellar beating results in the movement of fluid around them. This is the case in the ciliate epithelia which lines cavities in the body of animals, and many other examples can be found in the biological world. It is evident that motility depending on cilia and flagella can occur only in liquid environments. Amoeboid movement, on the contrary, is effective for the locomotion of cells in non-liquid environments. It is found not only in free-living micro-organisms but also in the higher living forms where it plays a crucial role during development and particularly in the morphogenetic arrangement of multicellular organisms (see *Cellular Development* by D. R. Garrod in this series). There are examples, however, in which the two types of movement are present at the same time in the same cell. More surprisingly, there are micro-organisms such as the soil amoeba *Naegleria* or the plasmodial slime moulds *Physarum* and *Didymium* that show an amoeboid-type locomotion when living on soil and develop flagella in response to water in the environment. Due to this characteristic, these micro-organisms may be regarded as linking forms between amoebae and flagellates.

Also, the movements within a cell are very common. This depends on the fact that the structural organization of a living cell is in a dynamic state and environmental conditions trigger responses that frequently necessitate intracellular movements. They may involve either the cell membrane, as in the process of phagocytosis, or the organelles within the cell, such as vacuoles, lysosomes, pigment granules etc., or the cytoplasm itself, e.g., during the cytoplasmic streaming (cyclosis) of plant cells.

Having defined in the preceding chapter the molecular components of the motility system in eukaryotic cells, in the following sections we will examine the mechanisms of flagellar and amoeboid movement and how it is controlled.

4.1 Ciliary and flagellar movement

Cilia and flagella of eukaryotic cells are cytoplasmic appendages able to generate movement. They consist basically of a microtubular axis (axoneme) covered by the cell membrane. Both appendages are structurally similar, the main difference being in the appearance of the movement, which consists of uniform undulations for flagella and biphasic beating (formed by a fast effective stroke and a slow recovery stroke) for cilia.

4.1.1 The mechanism of movement

According to the sliding filament model of Peter Satir [129], the movement of cilia and flagella is generated by the action of the dynein arms of microtubular doublets in the presence of ATP. The sliding movement thus originated is then transformed into bending the organelle by the presence of protein connections (radial spokes and nexin links) among doublets. The crucial role played by radial spokes and nexin links in sliding–bending conversion is shown by the fact that, when these connections are digested by trypsin, axonemes isolated from flagella or cilia lose their ability to propagate undulatory movements. Undigested axonemes, on the contrary, in the presence of ATP, bend in a way resembling intact flagella. From this simple experiment some important conclusions can be drawn. The first is that the axoneme itself is the portion of the flagellum responsible for movement. The second is that the movement is a consequence of the interaction between dynein arms and microtubule doublets. The third is that the protein connections between the axonemal doublets are responsible for the transformation of the sliding movement into a bending of the flagellum. The first two points have already been discussed in Chapter 3, Section 3.4, to which the reader is referred. Our interest will be focused therefore on the last item and particularly on how the sliding movement of adjacent doublets in the axoneme is transformed into bending of the flagellum. We shall also consider what the role of nexin links and radial spokes is in this transformation, and how local bending is able to produce a wave-like movement of the entire organelle.

Most of the present knowledge on the roles of nexin links and radial spokes comes from the work of Warner and Satir [168] on lateral cilia of molluscan gills (*Elliptio* spp.), which were fixed at different stages of the beat cycle and observed with the electron microscope, using a high resolution technique. By examining the behaviour of radial spokes in straight or unbent regions of the cilium, it has been shown that they are not linked to the central sheath and maintain a perpendicular orientation relative to the subfibre from which they originate. In bent regions, however, radial spokes are attached to the central sheath and show an angular displacement proportional to the curvature of the cilium. It seems, therefore, that radial spokes undergo an attachment-detachment cycle with respect to the central sheath according to the curvature of the cilium. In the areas where they are attached to the sheath in response to sliding, bending is generated. In the remaining parts of the cilium the unattached radial spokes allow free sliding of the microtubule doublets (Fig. 4.1). Excessive sliding is prevented by the nexin links connecting adjacent doublets. These structures seem not to undergo an

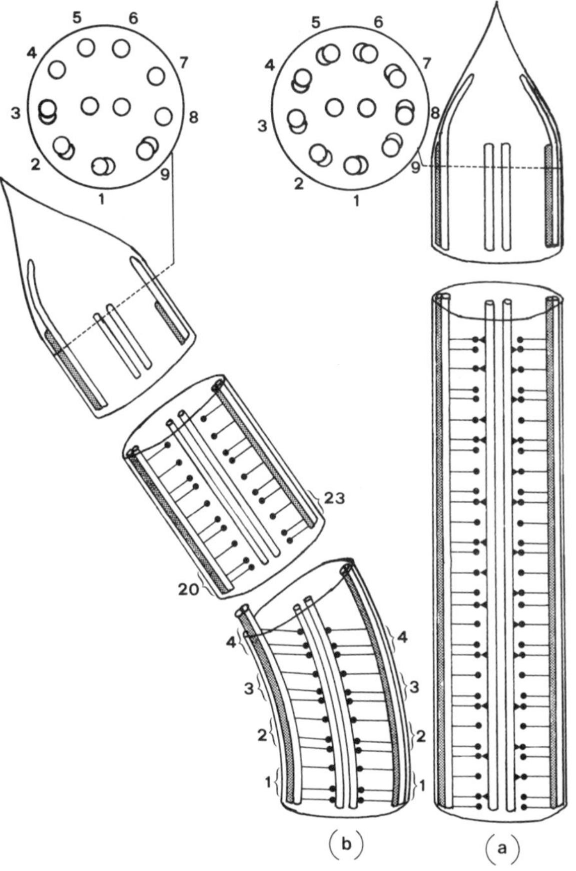

Fig. 4.1 A diagrammatic representation of a flagellar axoneme unbent (a) or during bending (b), showing the relationships between radial spokes and central doublets and displacement of doublets at flagellar tip in a cross-section.

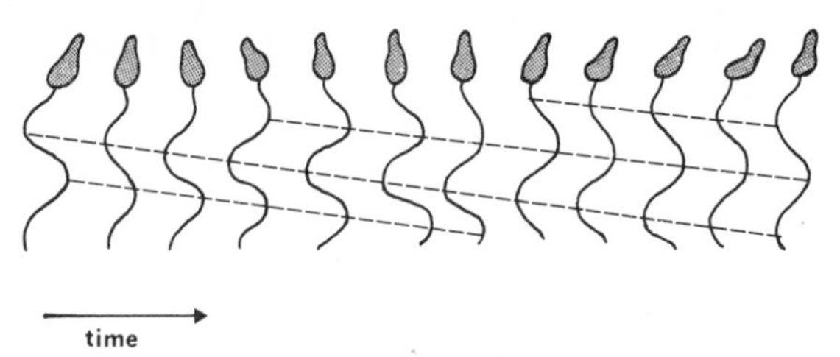

Fig. 4.2 Diagram of flagellar wave propagation during sperm movement with respect to time.

attachment–detachment cycle along the axoneme, but function as permanent elastic connections between doublets, limiting the amount of sliding that may occur.

4.1.2 Patterns of movement

In experimental conditions, the initiation of bending in cilia and flagella may occur at any point along the axoneme. This was demonstrated by Brokaw and Gibbons [26] using partially demembraned sperm flagella that, in the presence of ATP, formed bends only in those regions in which membranes has been removed. In natural conditions, however, a bend is most frequently initiated at the base of the flagellum or cilia and the wave propagates along it toward the tip (Fig. 4.2). There are also more rare exampes (as in the case of the protozoa trypanosomes) in which, during normal movement, waves pass along the flagellum from the tip toward the base.

Due to the favourable size of the structure, the patterns of movement of eukaryotic flagella can be easily studied with optical microscopy associated with techniques of image recording such as high-speed cinephotography. The movement is of an undulatory type and may take place in a single plane or in all the planes of the space. In the latter case the space into which the movement develops has a cylindrical appearance and the beat tends toward an elliptical form. Characteristic of the flagellar movement is the symmetry and the regularity of the waves. Since these originate at a remarkably fast rate (20–30 times per second), more than one wave is normally running through the flagellar body, and therefore the flagella appears in an undulated form (Fig. 4.3).

The movement of a flagellum in a liquid medium produces a resultant force able to propel a free-swimming organism in the direction opposite to that of the wave propagation. Therefore, in the case of base-to-tip wave propagation, the cell is pushed by the flagellum, and the reverse

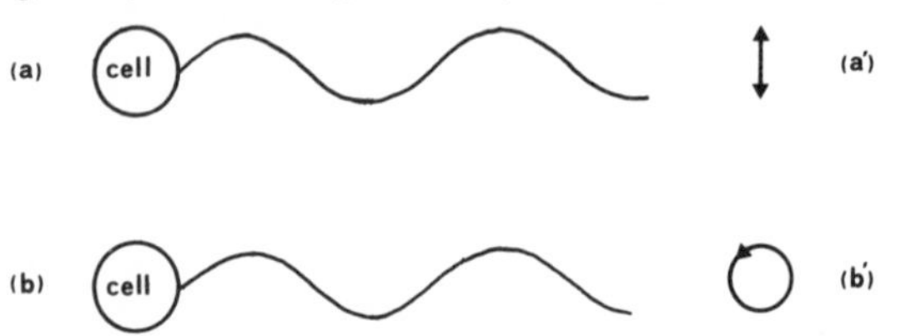

Fig. 4.3 The movement of flagella. In (a) planar movement; in (b) a helical movement; (a′) and (b′) show the course of movement of the corresponding tips.

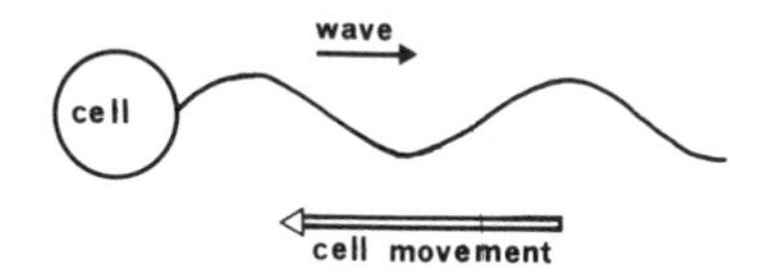

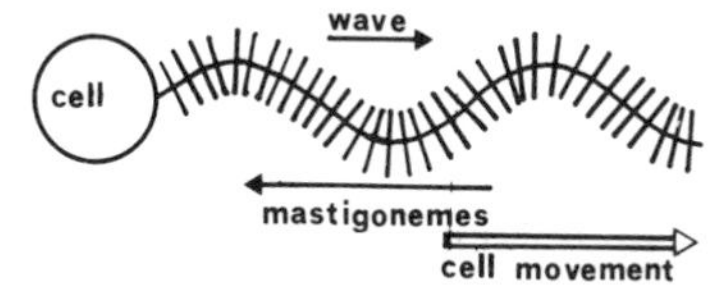

Fig. 4.4 Direction of cell movement as a consequence of flagellar motion in a cell without mastigonemes (top) and with mastigonemes (bottom).

occurs in tip-to-base wave propagation. An interesting exception to this assumption is shown by the *Ochromonad* flagellates, in which the movement of the organism follows the direction of the propagating wave. This is due to the fact that the flagellum of such protozoa possesses two rows of hairs (each about 20 nm thick and 1 μm long) called mastigonemes, which are attached on opposite sides along the entire organelle (Fig. 4.4). According to Jahn *et al.* [74], the flagellum beats in the plane containing the mastigonemes. These project rigidly form the flagellum and produce a flow of liquid in the direction opposite to that of the flagellar wave. Since the force generated by the mastigonemes is greater than that of the flagellar shaft, the propulsion of the organism will occur in the direction of the wave propagation.

The movement of cilia shows a great variety of patterns and is generally known as ciliary beating. According to a general model derived from the observation of ciliary beating in different organisms, it can be divided into two separate, sequential phases. During the first (effective stroke), the ciliary shaft bends in a region proximal to its base while the remaining part of the organelle remains straight or slightly curved. At the same time the ciliary tip describes an arc of at least 90°, lying on the same plane. The second phase (recovery stroke) is characterized by the propagation of a slow bend from the basal region toward the tip of the cilium which causes its return to the position held at the beginning of the effective stroke (Fig. 4.5). The effecive stroke is the most efficient phase of the ciliary beating: it produces a much greater movement of fluid than the recovery stroke and, therefore, determines the directionality of the movement. In case of free-living micro-or-

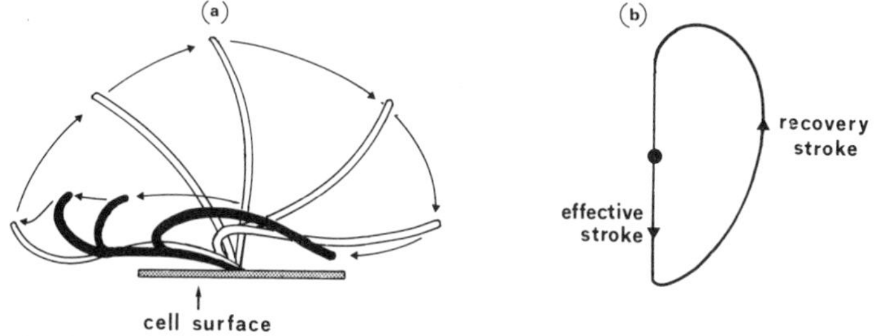

Fig. 4.5 A series of profiles showing the movement of a cilium during effective stroke and recovery stroke (a) and the path followed by ciliary tip (b).

ganisms, this consists of a cell movement opposite to the direction of the effective stroke. In ciliate epithelia or similar structures, the propulsion of fluids due to the ciliary beating is in the same direction as the effective stroke. During the recovery phase, the cilium slowly returns to the starting position occupying a plane different from that of the effective stroke. The slowness of the movement causes only minimal displacement of fluids, usually ineffective for motility purposes.

4.1.3 Coordination and control of movement

The coordination of movement of the motile appendages of eukaryotic cells is particularly evident in cilia. This is because cilia usually do not occur as single elements over a cell surface but, rather, form structures into which thousands of individual organelles are densely packed. In these conditions a more or less independent motion of each cilium is difficult to imagine and, in fact, ciliary beating in such structures occurs in a highly coordinated fashion. Two different ideas have been put forward to explain the mechanism of ciliary coordination. According to the first (neuroid coordination), motility is transmitted among cilia by a kind of neural mechanism in which the excitation of the ciliary shaft travels sequentially from one organelle to the other along cellular structures. The second type of coordination implies that coupling of movement between adjacent organelles results from mechanical forces due to the viscosity of the fluid into which cilia are moving (see Fig. 4.6). To distinguish between the two hypothesis, Michael Sleigh [138] has conducted a series of elegant experiments on *Pleurobranchia* comb plates using mechanical devices to stimulate or arrest ciliary movement. The result was that when a thin cellulose sheath, able to arrest the fluid motion, was inserted between adjacent cilia, the propagation of movement was immediately stopped. This experiment and other similar ones stress the importance of the movement of fluids for ciliary coordination and support the theory of mechanical coordination. It seems quite clear, however, that such a control does not operate in all the ciliate systems. In fact, there is evidence that in many metazoans cilia coordination depends on neuronal-type mechanisms.

When the surface of a ciliated cell is observed during ciliary movement, metachronal (out-of-phase) waves are seen animating the cell surface as a result of interciliary coordination. The image coming out of such a situation, most poetically, has been compared to 'waves of

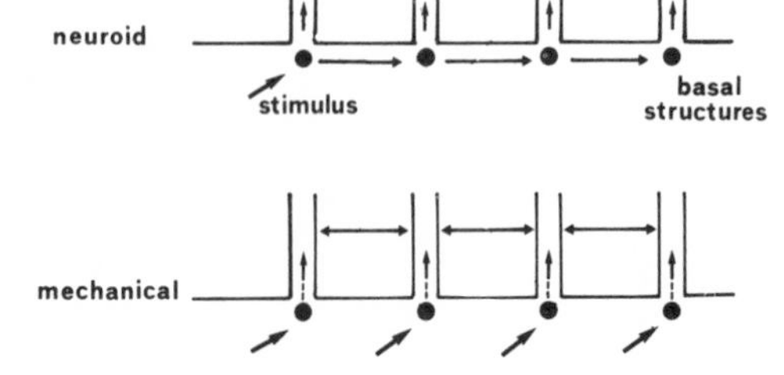

Fig. 4.6 Diagram illustrating a possible mechanism for neuroid and mechanical co-ordination and ciliary beating.

bending that pass over a field of corn blown by the wind'. The direction of the effective stroke of the ciliary beating and the direction of propagation of the metachronal waves in a field of organelles determine different patterns of coordinated ciliary beating that have been described in different organisms [77]. They are known as sympletic metachrony, when the metachronal wave and the effective stroke have both the same direction, anti-plectic metachrony, when the two directions are opposite, and diaplectic metachrony, when the directions lie at a right angle (Fig. 4.7). This pattern is specific for a given ciliate structure.

The control of the locomotion activity in flagellated or ciliated cells obviously depends on the possibility that the cell has to modify the motility behaviour of its appendages. These modifications, as in the case of bacterial motility, are most frequently triggered by environmental stimuli of various kinds, including chemical, electrical, mechanical or photonical stimuli. Environmental stimulants control, therefore, both the spatial orientation and the velocity of the cell locomotion.

One of the orientated reactions of eukaryotic micro-organisms studied in more detail is the so-called 'avoiding reaction' of flagellated and ciliated cells. This reaction was first interpreted as a mechanism for

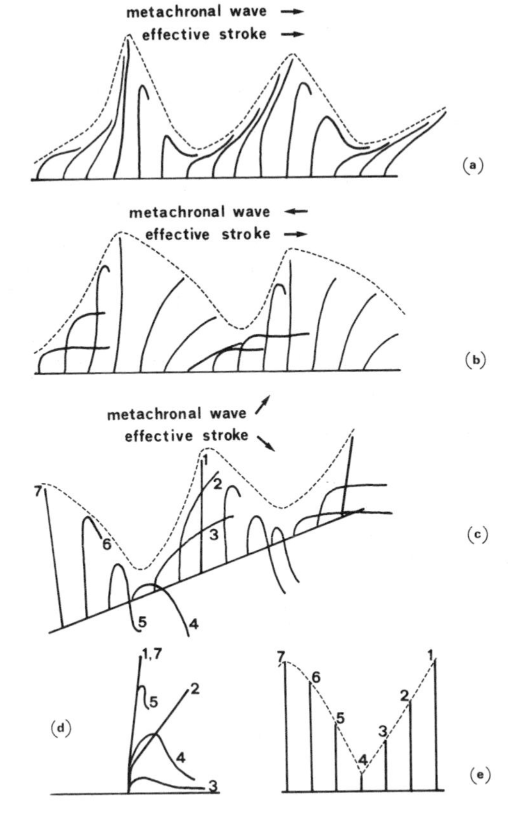

Fig. 4.7 Metachronal wave pattern. In (a) symplectic metachrony; in (b) anti-plectic; in (c), (d) and (e) dexioplectic; (d) shows cilia viewed in wave direction; (e) shows the appearance of the numbered cilia, in (c) viewed from a direction normal to wave propagation. (From Holwill, M.E.J., 1977, *Adv. Microbial. Physiol.*, **16**, 1–48.)

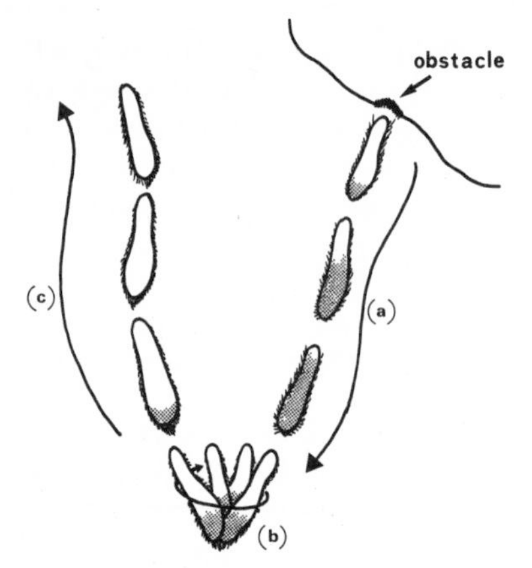

Fig. 4.8 Steps in the avoiding reaction of a *Paramecium* showing in (a) the escape from an obstacle, in (b) re-orientation of the cell, and in (c) movement in a new direction. (From [42].)

avoiding the mechanical obstacles found by a swimming cell on its way; subsequently, however, it has been shown to occur also in response to all the other tactic stimulants. In ciliates, such as *Paramecium*, a temporary reversal in the direction of the ciliary effective stroke causes a backward run to the cell. This reaction is evoked by mechanical or other physico-chemical stimulants. It ends when the cilia resume the normally directed power stroke and consequently the locomotion is again in the forward direction (Fig. 4.8). Extensive studies by Roger Eckert and his group have now clarified some of the aspects connected with the generation of the reversal of the ciliary beating by showing that it is controlled by the electrical potential of the cell membrane and by its interference with the permeability of the cell membrane to Ca^{2+} ions. (see [42, 102] for review). According to the currently accepted model, a stimulus at the anterior end of a *Paramecium* induces a depolarization of the cell membrane, causing an increase in Ca^{2+} conductance from outside to inside the cell. As a consequence, Ca^{2+} concentration within the cell increases and the cilia reverse their beating. If the depolarizing stimulus is not maintained (i. e., its duration is limited in time), Ca^{2+} is actively pumped out of the cell until its normal internal concentration is restored. When this point is reached, ciliary beating resumes the normal

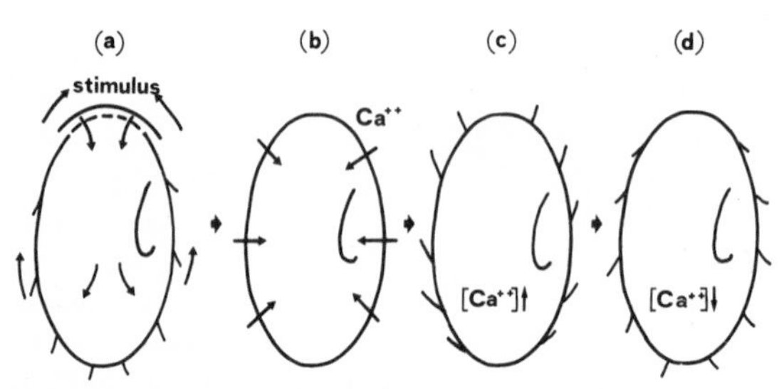

Fig. 4.9 The sequence of steps in the avoiding reaction showing in (a) membrane depolarization following stimulation, in (b) Ca^{2+} influx within the cell as a consequence of depolarization, in (c) Ca^{2+} accumulation and reversal of the ciliary beating, and in (d) resumption of the normal beating as a consequence of normal intracellular Ca^{2+} concentration. (From [42].)

orientation and the cell swims again in the forward direction (Fig. 4.9).

Apart from *Paramecium*, a Ca^{2+} control over the motile activity of cilia and flagella has been shown to operate also in spermatozoa of *Tubularia* or *Elliptio* and in flagellated protozoa such as *Crithidia oncopelti* or *Chlamydomonas*. How Ca^{2+} ions trigger the response of cilia and flagella, causing the reversal of beating is, at the moment, not completely understood.

The mechanism for the reorientation of the cell as a consequence of tactic stimulation is not identical in all micro-organisms. Biflagellates, such as *Chlamydomonas,* reverse the direction of swimming by extending the flagella before the cell for a short while. Some sperm cells, on the contrary, stop the flagellar motion momentarily and turn their head in a new direction, in which the swimming will then be resumed. In addition, trypanosomes change the direction of wave propagation along the flagellum from the normal tip-to-base direction to a new base-to-tip direction, therefore changing the directionality of the motion. All these different mechanisms are highly effective in the production of orientated motility and probably possess common mechanisms of generation.

4.2 Amoeboid movement

Among amoeboid movements are included all those forms of cell locomotion which depend on the production of transient cytoplasmic extensions of the cell surface, broadly defined as pseudopods. This form of movement has been originally defined in the amoeboid protozoa of the class *Sarcodina*, but is now clear that it is present in the majority of cells, although with some modifications. Pseudopodal extensions may be produced by a cell in all possible environments (water, air, soil, etc.). However, cell locomotion results only when the organism lies on a solid support. Therefore, only cells lying either on solid surfaces or within multicellular organisms (where the support is provided by other cells in the tissue) are capable of amoeboid movement.

4.2.1 The mechanism of movement

As previously discussed, amoeboid movement requires the presence of a system of microfilaments whose major component is actin. Actin filaments are able to interact with myosin in the presence of ATP and generate the propulsive force which results in movement. Although it is universally accepted that actomyosin sliding is the true motor for amoeboid movement, it is less clear how this motile activity is related to the pseudopodal formation and therefore to this type of locomotion.

Two major theories have been proposed to explain amoeboid movement. The first, *ectoplasmic tube contraction theory,* was developed by Pantin [109] and Mast [96] a long time before the demonstration of contractile microfilaments within the cell. According to this theory, a contraction of the peripheral, hyaline cytoplasmic area (known as cell ectoplasm) induces an increased hydraulic pressure on the more central endoplasm, pushing it towards the cell region of less pressure from which pseudopods originate (like squeezing a balloon, with the protrusion being the pseudopod) (Fig. 4.10a). The second *frontal zone contraction theory,* of Robert Allen [12], assumes a contraction of the endoplasm at the front of each advancing pseudopod able to pull part of the endoplasm forward (Fig. 4.10b). Although it has been shown that the ectoplasmic tube contraction is also consistent with the sliding filament model of acto-myosin interaction [123], at the moment Allen's hypothesis is the most widely accepted and has received much support both from experimental and theoretical data (see [14] for review, and [105]). It may be possible, however, that both mechanisms of pseudopodal formation are present in different systems.

But besides these theoretical considerations, the molecular mechanism of amoeboid movement is, at the moment, not sufficiently clear. Although some of the reactions which are involved in the process have been identified, much work still has to be done. The model proposed by Taylor and coworkers [154] seems, at this point, to fit quite nicely with the morphological and biochemical findings so far obtained. This model is based on the existence of a dynamic transition within the cell cytoplasm of monomeric actin into a filamentous state able to interact with myosin and generate contraction. After contraction has occurred, filamentous actin is again transformed into the monomeric form and the cycle is repeated. It is therefore a multi-step process involving the transformation of actin into a more structured state (gelation) followed by the contraction of the gel-like structure.

Within the cell, the zone of actin gelation is centred in the area of transition of endoplasm into ectoplasm. The relaxation of acto-myosin complexes, followed by the transformation of actin filaments into less structured states, may occur at the point of ectoplasm/endoplasm transition. This model stresses the importance of those factors able to control both the acto-myosin interaction and the process of gelation. It has been defined primarily using the large carnivorous amoeba *Amoeba proteus* as an experimental model, and verification of this model has been obtained in other systems.

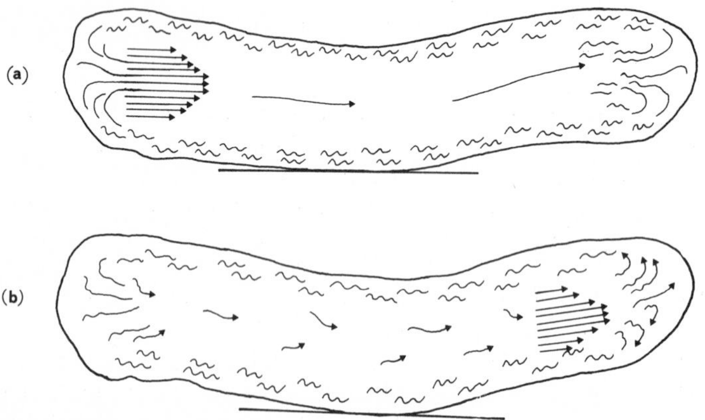

Fig. 4.10 Generation of force and endoplasmic streaming in amoeboid movement according to the (a) ectoplasmic tube contraction theory and (b) the frontal zone contraction theory.

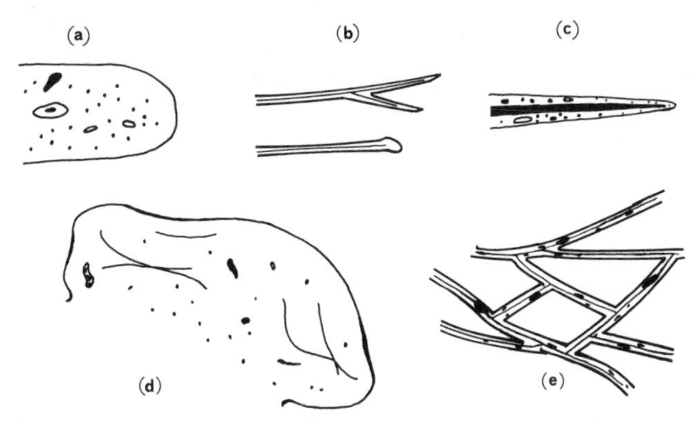

Fig. 4.11 The most frequently found forms of pseudopodia in cells: (a) = lobopodium; (b) = filopodia; (c) = axopodium; (d) = lamellipodium; (e) = reticulopodia.

4.2.2 Patterns of amoeboid locomotion

There are two requirements for a cell to be able to move in an amoeboid fashion. The first is the attachment to a solid substrate and the second is the emission of some kind of cytoplasmic appendage, known as a pseudopod, into which the rest of the cell eventually migrates. Although the basic pattern is as outlined, there is, however, considerable variation both in the shape and structure of the cytoplasmic appendages and in the pattern of locomotion in different cell types. From a morphological and structural point of view, at least five different forms of pseudopodia can be recognized (Fig. 4.11). Pseudopodia of the lobose type (or lobopodia) characteristic of *Amoebae* are broad curve-shaped appendages which can be seen originating from the cell membrane. They show at the tip a more transparent ectoplasmic area followed by a dense endoplasm which contains the cell organelles. Usually, during cell locomotion, lobopodia attach to a substrate and then the cell content flows into them, but in case of testate, or shelled amoebae, they act in a different way. These amoebae possess a shell of organic material inside of which is most of the cytoplasm. From the apertures of the shell extend finger-like lobopodia which make attachment with the substratum and contract, shortening and pulling the cell nearer to the attachment point.

Similar to lobopodia are the pseudopods of many types of mammalian cells known as lamellipodia. They are formed by a flattened sheet of cytoplasm, sometimes less than 0.1 μm thick, which extends into the anterior region of the cell. A lamellipodium contains microtubules, microfilaments and other cell organelles. It makes contact with the substrate in the advancing area, and drives the movement of the cell.

Filopodia are another type of pseudopodal extension found both in amoebae and in metazoan cells. They are very supple appendages (about 0.2 μm thick), consisting of an axis of microfilaments covered by the cell membrane. Filopodia are rapidly emitted and withdrawn by the cell body. They may function as exploratory appendages able to test the consistency of the surface before the cell migrates onto it [11].

Sometimes they transform into another type of pseudopodia (lobopodia and lamellipodia) which are more effective for locomotion purposes. Reticulopodia, mainly formed in *Foraminifera* protozoa, are slender appendages repeatedly branching and anastomosing in form of a fine network. They possess a microfilament axis and, in the cytoplasm surrounding it, many granules (mitichondria and cytoplasmic inclusions) can be seen. The entire reticulum has contractile properties and is used by the cell for movement. The last type of pseudopodal extension is the axopodia of the heliozoans which have been already described, both in term of structure and motile properties, in the section of microtubule movement (see page 41).

We shall now illustrate the amoeboid movement of the large soil amoebae *Chaos chaos* and *Amoeba proteus*, describing in some detail the more important morphological aspects of cell locomotion. The former are particularly suitable organisms for such kinds of studies, because of their body size which, being several millimetres long, greatly facilitates morphological studies. Mast [96] and more recently Allen [13] have provided accurate descriptions of the phenomenon. A moving amoeba is a polarized structure having an anterior pole corresponding to the pseudopodal tip and a posterior end, known as the tail or uroid. At the advancing front of the pseudopodal tip there is a thin hyaline cap of endoplasm, free of cytoplasmic granules, which encapsulates the cell. A granular endoplasm containing the nucleus and other cell organelles lies in the central region of the cell. In the advancing tip, the hyaline cap periodically expands forward by the abrupt transformation of part of the endoplasmic core of the pseudopod into a gel-like substance. During transformation, the pseudopod advances in the direction of locomotion at a maximum velocity of about 3–4 μm per second. The endoplasm streams toward the advancing pseudopod tip and the endoplasm which is immediately behind the cap diverges to either side, transforms into gel-like substance, and thereafter enters into the ectoplasmic layer. One can imagine this process being similar to the spurts of a fountain, and the area in which this occurs has therefore been called the 'fountain zone'. The ectoplasmic substance then flows

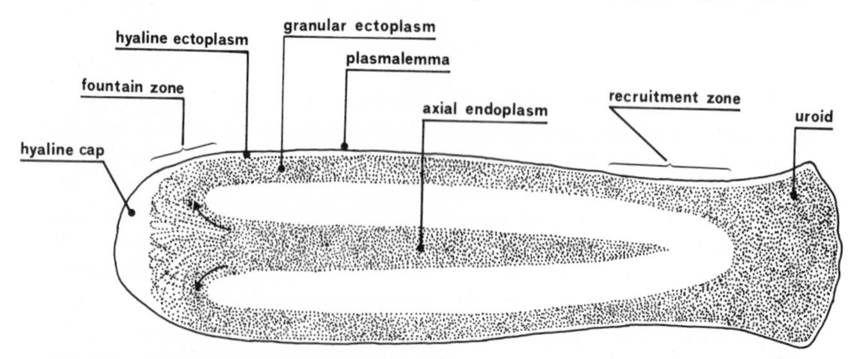

Fig. 4.12 Current terminology for different cytoplasmic regions in an amoeba. (From [13].)

backward towards the tail, or 'uroid region', where it again enters the endoplasm in a defined area called the 'recruitment zone'. As endoplasm, it again begins to move in the direction of the fountain zone.

Most important for amoeboid movement is the contact with the substrate on which the cell lies. In amoebae, the cell membrane of the central surface is attached to the substratum at a specific point. The portion of cell behind the attachment point shortens while the portion ahead lengthens, following the direction of movement.

The pattern of locomotion as seen in large amoebae cannot be generalized to all cells. It is likely, however, that all the amoeboid-type movements possess the same mechanism at a molecular level. What varies is mainly the gross morphology of the locomotion. In cultivated mammalian cells, such as fibroblasts, for instance, the advancing pseudopodium (or lamellipodium) has the form of a large lamella whose leading edge has a typical ruffling activity. The leading lamella moves forward and the rest of the cell follows. However, a fountain pattern of cytoplasmic streaming similar to that of a large amoebae cannot be seen. The cell has a terminating tail region, sometimes with a typical finger-like appearance. There is usually more than one site of adhesion between the ventral surface and the substrate which is distributed under lamellipodia, cell body and tail [1, 2].

4.2.3 The control of amoeboid locomotion

The ability to orientate cell movement in response to information of a chemical or physical nature coming from the environment is well developed in eukaryotic cells capable of amoeboid movements. In particular, chemotaxis (motility triggered by chemical stimuli) is exhibited by a variety of cells of higher organisms, such as fibroblasts, tumour cells, neurons, germ cells, leukocytes, and lower organisms such as amoeboid protozoa and slime moulds. However, although eukaryotic models for the study of orientated motility are quite common, chemotaxis has been studied in some detail only in polymorphonuclear leukocytes, among higher cells, and in the cellular slime mould *Dyctyostelium discoideum,* among lower eukaryotes. *D. discoideum,* in particular, provides excellent material for such studies, being easy to handle, allowing a multidisciplinary approach to the problem, and having a chemotactic system which is already fairly well defined.

(a) Chemotaxis in the cellular slime mould D. discoideum

D. discoideum, one of the most popular among cellular slime moulds, is a well-known model of cell differentiation. Its life cycle can be considered to be made up of three different phases: the growth phase, the aggregation phase, and the phase of morphogenetic development (Fig. 4.13). During the growth phase, the amoeboid cells grow and multiply using bacteria as a source of food. After the food is exhausted, the aggregation phase begins with the migration of amoebae towards centres where they form aggregates consisting of thousands of individual cells. These aggregates develop and differentiate further in the morphogenetic development phase, culminating

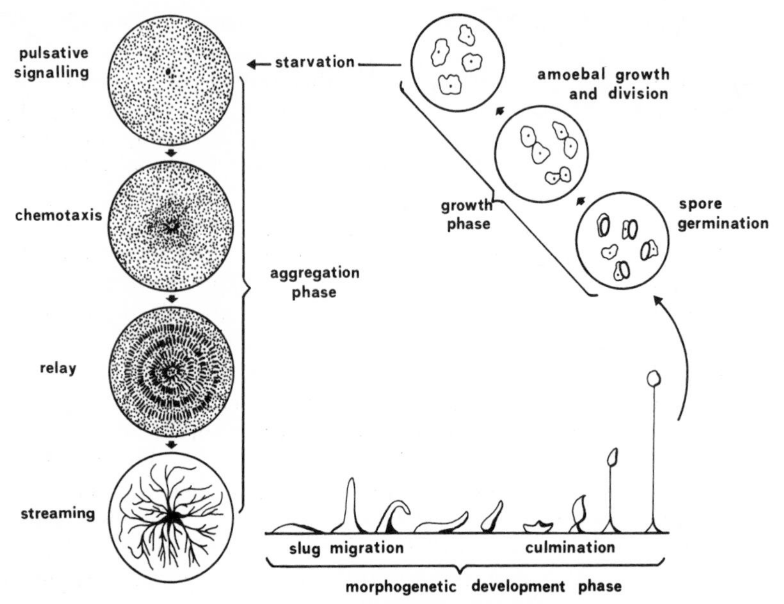

Fig. 4.13 The life cycle of *D.discoideum*. See text for details. (From P. C. Newell [10].)

with fruiting body construction in which three different cell types (basal disc, stalk cells, and spores) are evident. Under proper conditions spores eventually germinate into amoebae and the cell cycle is repeated. (For additional information see *Cell Differentiation* by J. M. Ashworth and reference [88].) The suggestion that chemotaxis was responsible for cell aggregation in *D. discoideum* had already been made at the beginning of the century by the first researchers slime moulds [106, 121]. It was then shown that the chemotactic substance (called acrasin) was a diffusible factor [126, 23] and this was partially characterized by Shaffer [131]. Finally, it was shown by Konjin and coworkers [79] that the nucleotide cyclic adenosine monophosphate (cAMP) was an attractant for the amoebae and that the natural chemotactic substance during aggregation of *D. discoideum* was cAMP. Cyclic adenosine monophosphate is also an attractant for other species of the same genus. Related slime moulds, however, (*Polyspondylium pollidum, P. violaceum*) utilize as an acrasin a small peptide with a molecular weight of about 1500 daltons. This is remarkable since leukocytes have a similar peptide chemo-attractant.

The aggregation process of *D. discoideum* is a complex phenomenon in which, using aggregation defective mutants, it has been calculated that about 200 genes are more or less directly involved. It starts with starvation that induces some of the amoebae to emit rhythmic pulses of cAMP. Production occurs at the rate of about 1 pulse every 10 min, and, as aggregation proceeds, increases to one pulse every 2-3 min. (Fig. 4.14). How pulse emission is regulated is not yet known. It may depend on a periodical change in activity of adenyl cyclase (the enzyme

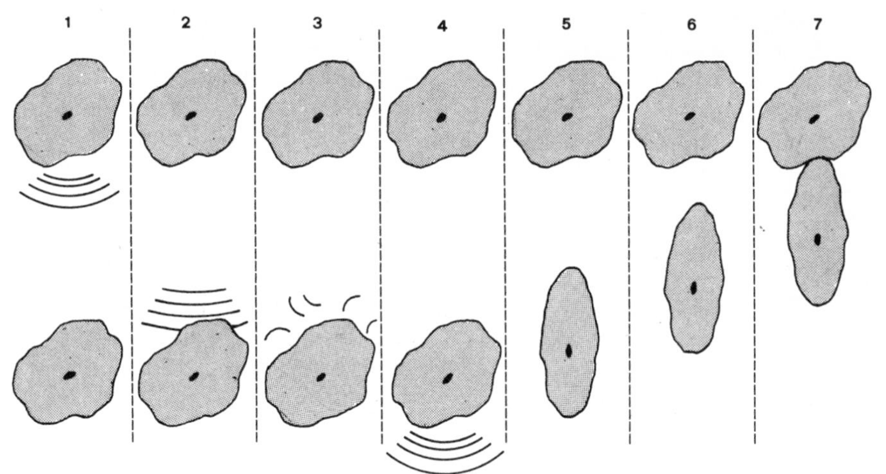

Fig. 4.14 The sequence of chemotactic signal and response in *D.discoideum* amoebae during aggregation. 1: signal emission, 2: signal reception, 3: signal destruction, 4: signal relay, 5, 6, 7: movement toward the signal.

that synthesizes cAMP) due to metabolic oscillations within the cell, but the problem is at the moment unresolved.

The cAMP diffuses around the cell from which it has been produced, reaching a maximum distance of about 100 μm. During its diffusion most of it is gradually destroyed by phosphodiesterases present in the environment but, before it completely disappears, some molecules bind to chemotactic receptors (cAMP receptors) located on the external surface of other amoebae. cAMP receptors are also produced as a response to starvation a few hours after the food is removed.

It is not clear how the receptors sense the chemotactic signal, and two possible explanations have been presented. According to the first, the receptors are able to sense a spatial gradient of the attractant, i. e., differences in concentration of cAMP in different parts of the cell. The second suggests that amoebae can sense changes in cAMP concentration with respect to time in a similar way to bacteria subjected to chemical stimuli.

The cAMP signal induces in the stimulated amoeba two types of time-limited responses: a motile response and the release of its own signal of cAMP. The motile response consists of the emission of a pseudopodium in the direction of the signal, resulting in the directed movement of the entire cell. Locomotion continues for about 100 sec, during which time the cell covers a distance of about 20 μm (corresponding to about twice the diameter of an amoeba). After this time, pseudopodia are withdrawn and cell movement stops until the cell receives a new chemotactic signal.

The second time-limited response is called the relay response and consists of the production and release of a pulse of cAMP about 12–15 sec after receiving the initial cAMP stimulus. During the relay response, there is an amplification of the stimulus since the cAMP pulse released can be as long as 60 sec. The periodicity of the signal emission and propagation within an aggregating field of amoebae induces the formation of aggregating circles of amoebae, each moving at the same time towards the centre, which may be followed easily with a microscope.

The process of aggregation continues then with the formation of streams of cells migrating towards an aggregation centre and ends when all the amoebae have reached the centre to which they were initially attracted (see [54] and [103]).

Such a model, in which the effect of chemotaxis on cell locomotion can be fully evaluated using both genetic and biochemical techniques, is very useful for studying microfilament-dependent cell motility. It is likely, therefore, that the results obtained in this system and in other similar ones may help to clarify all the events involved in chemotaxis and motility of eukaryotic cells in the fairly near future.

Topics for further reading

The various aspects of ciliary- and flagellar-dependent cell locomotion are specifically treated in:

Swimming and Flying in Nature (1975) (ed. T. Y. Wu, C. J. Brokaw and C. Brennen), Plenum Press, New York.

Cilia and Flagella (1975) (ed. M. Sleigh), Academic Press, London.

Eckert, R. (1972), Bioelectric control of ciliary activity, *Science*, **176**, 473–81.

Summers, K. (1975), The role of flagellar structures in motility, *Biochem. Biophys. Acta*, **416**, 153–68.

Holwill, M. E. (1977), Some biophysical aspects of ciliary and flagellar motility. *Adv. Microb. Physiol.*, **16**, 1–48.

Amoeboid locomotion is reviewed in more detail in:

Molecules and Cell Movements. A SPG symposium (1975) (ed. S. Inoué and R. Stephens), Raven Press, New York,

Which contains several reviews on amoeboid movement, and in many papers of the already mentioned:

Cell Motility (1976) (ed. R. Goldman, T. Pollard and J. Rosenbaum), Cold Spring Harbor Laboratory, Cold Spring Harbor.

Interesting articles are, in addition, in:

Allen, R. D. (1972), Biophysical aspects of pseudopodium formation, in *The Biology of Amoeba* (ed K. Jeon), *Academic Press, New York.*

Allen, D. A. and Stromger Allen, N. (1978), Cytoplasmic streaming in amoeboid movement. *Ann. Rev. Biophys. Bioeng.*, **7**, 469–95.

The reader who is curious about slime mould chemotaxis can pursue it further in the following articles:

Konijn, T. M. (1974), The chemotactic effect of cyclic AMP and its analogues in the acrasiae, *Antibiot. and Chemother.*, (Basel) **19**, 369–81.

Gerisch, G. and Malchow, D. (1976), Cyclic AMP receptors and the control of cell aggregation in *Dictyostelium*, in *Advances in Cyclic Nucleotide Research*, vol. 7 (ed P. Greengard and A. Robison), Raven Press, New York.

Bonner, J. T. (1977), Some aspects of chemotaxis using the cellular slime molds as an example, *Mycologia*, **69**, 443–59.

Newell, P. C. (1978), Cellular communication during aggregation of *Dictyostelium, J. Gen. Microbiol.*, **104**, 1–13.

References

[1] Abercrombie, M., Heaysman, J. and Pegrum, S. (1970), *Exp. Cell Res.*, **59**, 393–8.
[2] Abercrombie, M., Heaysman, J. and Pegrum, S. (1970), *Exp. Cell Res.*, **60**, 437–44.
[3] Abram, D. and Koffler, H. (1964), *J. Mol. Biol.*, **9**, 168–85.
[4] Adelstein, R. S. and Conti, M. A. (1975), *Nature*, **256**, 597–8.
[5] Adler, J. (1975), *Ann. Rev. Biochem.*, **44**, 341–56.
[6] Adler, J. and Dahl, H. H. (1967), *J. Gen. Microbiol.*, **46**, 161–73.
[7] Adler, J. and Templeton, B. (1967), *J. Gen. Microbiol.*, **46**, 175–84.
[8] Adler, J. and Epstein, W. (1974), *Proc. Nat. Acad. Sci. USA*, **71**, 2895–9.
[9] Afzelius, B. A. (1976), *Science*, **193**, 317–9.
[10] Albertini, D. F. and Clark, J. I. (1975), *Proc. Nat. Acad. Sci. USA*, **72**, 4976–80.
[11] Albrecht-Buehler, A. (1976), in *Cell Motility* (ed. R. Goldman, T. Pollard and J. Rosenbaum), Cold Spring Harbor Lab., pp. 247–64.
[12] Allen, R. D. (1961), *Exp. Cell Res.*, **8** (suppl.), 17–31.
[13] Allen, R. D. (1961), in *The Cell* (ed. J. Brachet and A. E. Mirsky) vol. 2, Academic Press, New York, pp. 135–216.
[14] Allen, R. D. and Stromgren Allen, N. (1978), *Ann. Rev. Biophys. Bioeng.*, **7**, 469–95.
[15] Aksamit, R. R. and Koshland, D. E. (1974), *Biochemistry*, **13**, 4473–8.
[16] Amstrong, J. B. (1972), *Can. J. Microbiol.*, **18**, 1695–1703.
[17] Arai, T. and Kaziro, Y. (1977), *J. Biochem.* (Tokyo), **82**, 1063–71.
[18] Barra, H. S., Arce, C. A., Rodriguez, J. A. and Caputto, R. (1974), *Biochem. Biophys. Res. Comm.*, **60**, 1384–90.
[19] Berg, H. C. (1971), *Rev. Sci. Instrum.*, **42**, 868–71.
[20] Berg, H. C. and Anderson, R. A. (1973), *Nature*, **245**, 380–2.
[21] Berry, R. W. and Shelansky, M. L. (1972), *J. Mol. Biol.*, **71**, 71–80.
[22] Bhisey, A. and Freed, J. (1971), *Exp. Cell Res.*, **64**, 419–29.
[23] Bonner, J. T. (1947), *J. Exp. Zool.*, **106**, 1–26.
[24] Borisy, G. G., Olmsted, J. B., Marcum, J. M. and Allen, C. (1974), *Fed. Proc.*, **33**, 167–74.
[25] Boss, W., Gordon, A. S., Hall, L. E. and Price, D. (1972), *J. Biol. Chem.*, **247**, 917–24.
[26] Brokaw, C. J. and Gibbons, I. K. (1973), *J. Cell Sci.*, **13**, 1–18.
[27] Bryan, J. and Wilson, C. (1971), *Proc. Nat. Acad. Sci. USA*, **68**, 1762–6.
[28] Bryan, R. N., Cutter, G. A. and Hayashi, M. (1978), *Nature*, **272**, 81–3.
[29] Cappuccinelli, P., Cuccureddu, R. and Hames, B. D. (1977), in *Development and Differentiation in the Cellular Slime Moulds* (ed. P. Cappuccinelli and J. M. Ashworth), Elsevier/North Holland, Amsterdam, pp. 231–41.
[30] Cappuccinelli, P., Martinotti, G. and Hames, B. D. (1978), *FEBS Letters*, **91**, 153–7.
[31] Chang, J. Y., De Lange, R. J., Shaper, J. M. and Glazer, A. N. (1976), *J. Biol. Chem.*, **251**, 695–700.

[32] Clarke, M. and Spudich, J. A. (1977), *Ann. Rev. Biochem.*, **46**, 797–822.
[33] Dales, S. (1972), *J. Cell Biol.*, **52**, 748–52.
[34] David-Pfeuty, T., Laparte, J. and Pantaloni, D. (1978), *Nature*, **272**, 282–4.
[35] De Pamphilis, M. L. and Adler, J. (1971), *J. Bacteriol.*, **105**, 384–95.
[36] De Pamphilis, M. L. and Adler, J. (1971), *J. Bacteriol.*, **105**, 396–407.
[37] De Rosier, D., Mandelkow, E., Silliman, A., Tilney, L. and Kane, R. (1977), *J. Mol. Biol.*, **113**, 679–95.
[38] Dimmit, K. and Simon, M. (1971), *J. Bacteriol.*, **105**, 369–75.
[39] Dimmit, K. and Simon, M. (1971), *J. Bacteriol.*, **108**, 285–6.
[40] De Petris, S. (1975), *J. Cell Biol.*, **65**, 123–46.
[41] Ebashi, S. and Endo, M. (1960), *Prog. Biophys. Mol. Biol.*, **18**, 123–83.
[42] Eckert, R. (1972), *Science* **176**, 476–81.
[43] Eipper, B. A. (1972), *Proc. Nat. Acad. Sci. USA*, **69**, 2283–7.
[44] Emerson, S. U., Tokynasu, K. and Simon, M. I. (1970), *Science*, **169**, 190–2.
[45] Felix, H. and Strauli, P. (1976), *Nature*, **261**, 603–5.
[46] Flanagan, J. and Koch, G. L. E. (1978), *Nature*, **273**, 278–81.
[47] Frére J. M. (1977), *Bull. Inst. Pasteur*, **75**, 187–203.
[48] Fujiwara, K. and Pollard, T. D. (1978), *J. Cell Biol.*, **67**, 125a.
[49] Fujiwara, K. and Pollard, T. D. (1978), *J. Cell Biol.*, **77**, 188–95.
[50] Fulton, C. and Simpson, P. M. (1976), in *Cell Motility* (ed. R. Goldman, T. Pollard and J. Rosenbaum), Cold Spring Harbor Lab., pp. 987–1006.
[51] Fuller, G. M., Brinkley, B. R. and Boughter, M. J. (1975), *Science*, **187**, 948–50.
[52] Gaskin, F. and Shelanski, M. L. (1976), in *Essays in Biochemistry* (ed. P. N. Campbell and W. N. Aldridge), vol. 12, Academic Press, New York, pp. 115–46.
[53] Gaskin, F., Cantor, C. R., and Shelansky, M. L. (1974), *J. Mol. Biol.*, **89**, 737–58.
[54] Gerish, G., Malchow, D. and Hess, J. (1974), in *Biochemistry of Sensory Functions, Mosbacher Colloquium* (ed. L. Jaenicke), Springer-Verlag, Berlin, pp. 279–98.
[55] Gibbons, B. H. and Gibbons, I. R. (1972), *J. Cell Biol.*, **54**, 75–97.
[56] Gibbons, I. R. (1965), *Arch. Biol*, **76**, 317–24.
[57] Gibbons, I. R., Fronk, E., Gibbons, B. and Ogawa, K. (1976), in *Cell Motility* (ed. R. Goldman, T. Pollard and J. Rosenbaum), Cold Spring Harbor Lab., pp. 905–32.
[58] Goldman, R. (1971). *J. Cell Biol.*, **51**, 752–62.
[59] Goldman, R. and Follet, E. (1970), *Science*, **169**, 268–88.
[60] Gragolev, A. N. and Skulachev, V. P. (1978), *Nature*, **272**, 380–2.
[61] Grasse, P. P. (1956), *Arch. Biol.*, **67**, 595–609.
[62] Grimstone, A. V. and Cleveland, L. R. (1965), *J. Cell Biol.*, **24**, 387–400.
[63] Hazelbauer, G. L. and Adler, J. (1971), *Nature New Biol.*, **230**, 101–4.
[64] Herzhog, W. and Weber, K. (1977), *Proc. Nat. Acad. Sci. USA*, **74**, 1860–4.
[65] Hilmer, M. and Simon, M., in *Cell Motility* (ed. R. Goldman, T. Pollard and J. Rosenbaum), Cold Spring Harbor Lab., pp. 35–45.
[66] Himes, R. H., Burton, P. R. and Gaito, J. M. (1977), *J. Biol. Chem.*, **252**, 6222–8.
[67] Hitchen, E. T. and Butler, R. D. (1973), *Z. Zellforsch. mikrosk. anat.*, **144**, 59–73.
[68] Hitchcock, S., Carlsson, L. and Lindberg, U. (1976), in *Cell Motility* (ed. R. Goldman, T. Pollard and J. Rosenbaum), Cold Spring Harbor Lab., pp. 545–59.

[69] Hollande, A. and Gharagzoglu, I. (1967), *C. R. Acad. Sci.*, **265**, 1309–15.
[70] Hunter, T. and Garrels, J. I. (1972), *Cell*, **12**, 767–81.
[71] Iino, T. (1977), *Ann. Rev. Genet.*, **11**, 161–82.
[72] Iino, T. and Mitani, M. (1967), *J. Gen. Microbiol.*, **49**, 81–8.
[73] Inoué, S. and Ritter, N. (1975), in *Molecules and Cell Movement* (ed. S. Inoué and R. E. Stephens), Raven Press, New York, pp. 3–30.
[74] Jahn, T. L., Landman, M. D. and Fonseca, J. R. (1964), *J. Protozool.*, **11**, 291–303.
[75] Jakus, M. A. and Hall, C. E. (1946), *Biol. Bull.*, **91**, 141–4.
[76] Joys, T. M. and Rankis, V. (1972), *J. Biol. Chem.*, **247**, 5180–93.
[77] Knight-Jones, E. W. (1954), *Q. J. Microsc. Sci.*, **95**, 503–21.
[78] Kondok, H. and Yanagida, M. (1975), *J. Mol. Biol.*, **96**, 641–52.
[79] Konijn, T. M., Barckley, D. S., Chang, Y. Y. and Bonner, J. T. (1968), *Amer. Naturalist*, **102**, 225, 234.
[80] Korn, E. D. (1978), *Proc. Nat. Acad. Sci. USA*, **75**, 588–99.
[81] Kowitt, J. D. and Fulton, C. (1974), *J. Biol. Chem.*, **249**, 3638–46.
[82] Kowitt, J. D. and Fulton, C. (1974), *Proc. Nat. Acad. Sci. USA*, **71**, 2877–81.
[83] Larsen S. H., Adler, J., Gargus, J. J. and Hogg, R. W. (1974), *Proc. Nat. Acad. Sci. USA*, **72**, 1239–43.
[84] Lazarides, E. (1976), in *Cell Motility* (ed. R. Goldman, T. Pollard, and J. Rosenbaum), Cold Spring Harbor Lab., pp. 347–60.
[85] Lazarides, E. and Lindberg, U. (1974), *Proc. Nat. Acad. Sci. USA*, **71**, 4742–6.
[86] Lazarides, E. and Weber, K. (1974), *Proc. Nat. Acad. Sci. USA*, **71**, 2268–72.
[87] Linck, R. W. (1976), *J. Cell Sci.*, **20**, 405–39.
[88] Loomis, W. F. (1975), *Dictyostelium Discoideum: a Developmental System*, Academic Press, New York.
[89] Luduena, R. F. and Woodward, D. O. (1973), *Proc. Nat. Acad. Sci. USA*, **70**, 3594–8.
[90] MacNab, R. and Koshland Jr., D. E. (1972), *Proc. Nat. Acad. Sci. USA*, **69**, 2509–12.
[91] MacNab, R. and Koshland Jr., D. E. (1974), *J. Mol. Biol.*, **84**, 399–406.
[92] MacNab, R. and Orston, M. K. (1977), *J. Mol. Biol.*, **112**, 1–30.
[93] Manson, M. D., Tedesco, P., Berg, H. C., Harold, F. M. and Van der Drift, C. (1977), *Proc. Nat. Acad. Sci. USA*, **74**, 3060–4.
[94] Margulis, L., To, L., and Chase, D. (1978), *Science*, **200**, 1118–24.
[95] Maruta, H. and Korn, E. D. (1977), *J. Biol. Chem.*, **252**, 8329–32.
[96] Mast, S. D. (1925), *J. Morphol. Physiol.*, **41**, 347–425.
[97] Morgan J. L., and Seeds, N. W. (1975), *J. Cell Biol.*, **67**, 136–45.
[98] Mooseker, M. S. (1976), *J. Cell Biol.*, **71**, 417–33.
[99] Mooseker, M. S. and Tilney, L. G. (1973), *J. Cell Biol.*, **56**, 13–26.
100] Mooseker, M. S. and Tilney, L. G. (1976), *J. Cell Biol.*, **67**, 724–43.
101] Murphy, D. B. and Borisy, G. C. (1975), *Proc. Nat. Acad. Sci. USA*, **72**, 2696–700.
102] Naitoh, Y. and Eckert, R. (1975), in *Cilia and Flagella* (ed. M. A. Sleigh), Academic Press, New York, pp. 305–55.
103] Newell, P. C. (1978), *J. Gen. Microbiol.*, **104**, 1–13.
104] Nicolson, G. L. (1976), *Biochem. Biophys. Acta*, **457**, 57–108.
105] Odell, G. M. and Frisch, H. L. (1975), *J. Theor. Biol.*, **50**, 59–86.
106] Olive, E. W. (1902), *Proc. Boston Soc. Nat. Hist.*, **30**, 451–513.
107] Olmsted, J. B. and Borisy, G. G. (1973), *Ann. Rev. Biochem.*, **42**, 507–40.

[108] Olmsted, J. B. and Borisy, G. G. (1975), *Biochemistry*, **14**, 2996–3005.
[109] Pantin, C. F. A. (1923), *J. Marine Biol. Ass.*, **13**, 24–69.
[110] Pastan, I. and Perlman, R. (1970), *Science*, **169**, 339–42.
[111] Pedersen, H. and Rebbe, H. (1975), *Biol. Reprod.*, **12**, 541–4.
[112] Penningroth, S. M., Cleveland, D. M. and Kirschner, M. W. (1976), in *Cell Motility* (ed. R. Goldman, T. Pollard and J. Rosenbaum), Cold Spring Harbor Lab., pp. 1233–55.
[113] Pickett-Heaps, J. D. (1969), *Cytobios*, **1**, 257–80.
[114] Piperno, G. and Luck, D. (1974), *J. Cell Biol.*, **63**, 271a.
[115] Piperno, G. and Luck, D. J. (1977), *Biol. Chem.*, **252**, 383–91.
[116] Piperno, G., Huang, B. and Luck, D. J. (1977), *Proc. Nat. Acad. Sci. USA*, **74**, 1600–4.
[117] Pollard, T. D. and Korn, E. D. (1973), *J. Biol. Chem.*, **248**, 448–54.
[118] Pollard, T. D. and Korn, E. D. (1973), *J. Biol. Chem.*, **248**, 4682–90.
[119] Pollard, T. D. and Korn, E. D. (1973), *J. Biol. Chem.*, **248**, 4691–7.
[120] Pollard, T. D., Porter, N. E. and Stafford, D. W. (1977), *Cell Biol.*, **75**, 262a.
[121] Potts, G. (1902), *Flora*, **91**, 281–347.
[122] Raybin, D. and Flavin, M. (1975), *J. Cell Biol.*, **67**, 356a.
[123] Rinaldi, R., Opas, M. and Hrebenda, B. (1975), *J. Protozool.*, **22**, 286–92.
[124] Robbins, E. and Shelanski, M. L. (1969), *J. Cell Biol.*, **43**, 371–3.
[125] Robinson, W. G. (1972), *J. Cell Biol.*, **52**, 66–83.
[126] Runyon, E. H. (1942), *Collecting Nat.*, **17**, 99–129.
[127] Sabatini, D., Bensch, K., Barnett, R. (1963), *J. Cell Biol.*, **17**, 19–58.
[128] Satir, P. (1965), *J. Cell Biol.*, **26**, 805–34.
[129] Satir, P. (1968), *J. Cell Biol.*, **39**, 77–94.
[130] Satir, P. (1974), *Sci. Amer.*, **231**, 44–54.
[131] Shaffer, B. M. (1956), *J. Exp. Biol.*, **33**, 645–57.
[132] Sheir-Neiss, G., Nardi, R. V., Gealt, M. A. and Morris, N. R. (1976), *Biochem. Biophys. Res. Comm.*, **69**, 285–90.
[133] Shelansky, M. L., Gaskin, F. and Cantor, C. R. (1973), *Proc. Nat. Acad. Sci. USA*, **70**, 765–8.
[134] Silverman, M. and Simon, M. (1974), *Nature*, **249**, 73–4.
[135] Silverman, M. and Simon, M. (1977), *Ann. Rev. Microbiol.*, **31**, 397–419.
[136] Silverman, M. and Simon, M. (1977), *Proc. Nat. Acad. Sci. USA*, **74**, 3317–21.
[137] Simon, M., Silverman, M, Latsumura, P., Ridgway, H., Komeda, Y. and Hilman, M. (1978), in *Relations between Structure and Function in the Prokaryotic Cell* (ed. R. Y. Stayner, H. J. Rogers and B. J. Ward), Cambridge University Press, pp. 272–87.
[138] Sleigh, M. A. (1972), in *Essay in Hydrobyology* (ed. R. B. Clark and R. Wootons), University of Exeter, pp. 119–36.
[139] Sleigh, M. A. ed. (1974), *Cilia and Flagella*, Academic Press, London.
[140] Sloboda, R. D., Rudolph, S. A., Rosenbaum, J. L. and Greengard, P. (1975), *Proc. Nat. Acad. Sci. USA*, **72**, 177–81.
[141] Sloboda, R. D., Dentler, W. L., Bloodgood, R. A., Telzer, B. R., Granett, S. and Rosenbaum, J. L. (1976), in *Cell Motility* (ed. R. Goldman, T. Pollard and J. Rosenbaum), Cold Spring Harbor Lab., pp. 1171–212.
[142] Snyder, J. A. and McIntosh, J. R. (1976), *Ann. Rev. Biochem.*, **45**, 699–720.
[143] Springer, M. J., Goy, M. F. and Adler, J. (1977), *Proc. Nat. Acad. Sci. USA*, **74**, 3312–6.
[144] Spudich, J. A. and Watt, S. (1971), *J. Biol. Chem.*, **246**, 4866–71.

[145] Starger, G. M. and Goldman, R. D. (1977), *Proc. Nat. Acad. Sci. USA*, **74**, 2422–6.
[146] Stephens, R. E. (1970), *Biol. Bull.*, **139**, 438a.
[147] Stephens, R. E. (1975), *J. Cell Biol.*, **67**, 418a.
[148] Stephens, R. E. (1975), in *Molecules and Cell Movements* (ed. S. Inoué and R. E. Stephens), Raven Press, New York, pp. 181–206.
[149] Stephens, R. E. and Edds, K. T. (1976), *Physiol. Rev.*, **56**, 709–77.
[150] Stephens, R. E., Renaud, F. L. and Gibbons, I. L. (1967), *Science*, **156**, 1606–8.
[151] Storti, R. V., Coen, D. M. and Rich, D. (1976), *Cell*, **8**, 521–7.
[152] Summers, K. E. and Gibbons, I. R. (1977), *Proc. Nat. Acad. Sci. USA*, **68**, 3092–6.
[153] Sundquist, K. G. and Ehrust, A. (1976), *Nature*, **264**, 4866–71.
[154] Taylor, D. L. (1976), in *Cell Motility* (ed. R. Goldman, T. Pollard and J. Rosenbaum), Cold Spring Harbor Lab., pp. 797–821.
[155] Taylor, B. and Koshland Jr., D. E. (1974), *J. Bacteriol.*, **119**, 640–2.
[156] Taylor, D. L., Rhodes, J. A. and Hammond, S. A. (1976), *J. Cell Biol.*, **70**, 123–43.
[157] Tilney, L. G. (1975), in *Molecules and Cell Movements* (ed. S. Inoue and R. E. Stephens), Raven Press, New York, pp. 339–88.
[158] Tilney, L. G. (1976), *J. Cell Biol.*, **69**, 73–89.
[159] Tilney, L. G. (1977), in *International Cell Biology* (ed. B. B. Brinkley and K. R. Porter), The Rockefeller University Press, New York, pp. 388–402.
[160] Tilney, L. G. (1978), *J. Cell Biol.*, **77**, 5051–64.
[161] Tilney, L. G. and Byers, B. (1969), *J. Cell Biol.*, **43**, 148–65.
[162] Tsang N., MacNab, R. and Koshland Jr., D. E. (1973), *Science*, **181**, 60–3.
[163] Tucker, J. B. (1968), *J. Cell Sci.*, **3**, 493–514.
[164] Tucker, J. B. (1974), *J. Cell Biol.*, **62**, 424–37.
[165] Tucker, J. B. (1977), *Nature*, **266**, 22–6.
[166] Vasiliev, J. M. and Gelfand, I. M. (1976), in *Cell Motility* (ed. R. Goldman, T. Pollard and J. Rosenbaum), Cold Spring Harbor Lab., pp. 279–304.
[167] Walters, B. B. and Mathus, A. I. (1975), *Nature*, **257**, 496–8.
[168] Warner, F. D. and Satir, P. (1974), *J. Cell Biol.*, **63**, 35–63.
[169] Watters, C. (1968), *J. Cell Sci.*, **3**, 231–44.
[170] Water, R. D. and Kleinsmith, L. J. (1976), *Biochem. Biophys. Res. Commun.*, **70**, 704–8.
[171] Weber, K. (1976), in *Cell Motility*, (ed. R. Goldman, T. Pollard and J. Rosenbaum), Cold Spring Harbor Lab., pp. 403–17.
[172] Weber, K., Bibring, T. and Osborn, M. (1975), *Proc. Nat. Acad. Sci. USA*, **72**, 459–63.
[173] Weingarten, M. B., Lockweed, A. H., Hwo, S. Y. and Kirschnar, M. W. (1975), *Proc. Nat. Acad. Sci. USA*, **72**, 1858–62.
[174] Weisenberg, R. C. (1972), *Science*, **177**, 1104–5.
[175] Wilson, L. and Bryan, J. (1974), *Adv. Cell Mol. Biol.*, **3**, 21–72.
[176] Wilson, L., Creswell, M. and Chin, D. (1975), *Biochemistry*, **14**, 5586–92.
[177] Witman, G. B., Plummer, J. and Sander, G. (1978), *J. Cell Biol.*, **76**, 729–47.
[178] Yamaguchi, S., Iino, T. and Kurdiwa, T. (1972), *J. Gen. Microbiol.*, **70**, 299–303.
[179] Zukin, R. S. and Koshland, Jr. D. E. (1976), *Science*, **193**, 405–8.

Index